The Science of
Structural
Engineering

The Science of
Structural
Engineering

Jacques Heyman
Emeritus Professor of Engineering, University of Cambridge

Imperial College Press

Published by

Imperial College Press
57 Shelton Street
Covent Garden
London WC2H 9HE

Distributed by

World Scientific Publishing Co. Pte. Ltd.
5 Toh Tuck Link, Singapore 596224
USA office: 27 Warren Street, Suite 401-402, Hackensack, NJ 07601
UK office: 57 Shelton Street, Covent Garden, London WC2H 9HE

Library of Congress Cataloging-in-Publication Data
Heyman, Jacques.
    The science of structural engineering / Jacques Heyman.
      p. cm.
    ISBN 1-86094-189-3 (pbk.)
    1. Structural engineering. I. Title.
  TA633.H49   1999
  624.1--dc21   99-34732
                            CIP

British Library Cataloguing-in-Publication Data
A catalogue record for this book is available from the British Library.

First published 1999
Reprinted 2006

Printed by FuIsland Offset Printing (S) Pte Ltd, Singapore

Preface

The design of a wooden table has, traditionally, been in the hands of crafts-men. Fashion may cause changes in style, but the legs of the table, whatever their decorative shape, will have sufficiently robust dimensions that they will stand up to normal use – which may well include someone sitting or standing on the table top. There is no place for advanced mathematics in the design of such a table, although this will not necessarily deter scientists or engineers (and the distinction between the two professions is remarked in Chapter 1 of this book) from attempting to determine how the forces are carried from the table top to the ground. And in fact such an analysis will be useful if the traditional wooden material is replaced by say plastic or lightweight metal – the legs of the table may then have to be designed carefully if they are to carry their loads satisfactorily.

The design of the legs of the table may be taken as a simple but archety-pal problem for the structural engineer – first, the forces to be carried by the legs must be calculated, and then the legs must be proportioned so that those forces are carried comfortably. It turns out that both these steps in design can be difficult. The four-legged table is, in technical jargon, hyper-static; a table supported by three legs is easy to analyse, but the addition of a fourth leg greatly complicates the problem. The way in which the weight of a man, standing at a specified off-centre point on the table, is distributed to the four legs cannot be determined without a complex scheme of calcu-lation. Such schemes of calculation form the subject matter of the *theory of structures* – the later step, the actual proportioning of the legs so that they are able to carry their loads, forms part of the science of *strength of materials*. The two steps cannot, in practice, always be separated.

There is, moreover, a complicating factor which may be appreciated by anyone who has dined outside a restaurant on a hot summer's night, with the portable table placed on the pavement. The table, annoyingly (both for the users and for the structural analyst), rocks. At any particular time a leg may be off the ground, and hence cannot be carrying any load – the other three legs must support the total weight of the table and its contents. Moreover, a passing waiter may nudge the table so that it takes up a new position, with a different leg off the ground. An attentive waiter will slice a wine cork lengthways on the slant to form two wedges, one of which may be put under an offending leg – that leg will then be supported not on hard pavement, but on a flexible footing. How, then, are the design values of the forces in the legs to be assigned, if any one of the legs may be carrying no load, or be supported rigidly or by a soft foundation?

The answer to this question is to be found in the application of the so-called plastic theory, which is discussed in Chapter 7, the final chapter of this book. Plastic theory was developed in this century, and it provides a way of designing a structure (like the four-legged table) whose precise state cannot be specified. All structures, except the very simplest, are of this kind and, in the past, designers have made calculations for conceptual models which ignore this inconvenient fact. As a result, those calculations lead to values – of stresses, for example – which cannot be observed in the real construction.

The seven sections of this book give accounts, in non-mathematical terms, of various aspects of structural theory. The presentation is roughly chronological, and, in a sense, forms a skeletal history of the subject. However, the objective is not to rake over the past but to illuminate the activity of the present-day structural engineer, and to show how a store of scientific information can be used creatively in design.

[No detailed bibliography is given. It is of interest that, for example, Euler solved the buckling problem in 1744, and this date is noted; only a dedicated scholar, however, will wish to consult the original Latin text. A full list of historical references may be found in J Heyman, *Structural Analysis: A Historical Approach*, Cambridge University Press, 1998; similarly *The Stone Skeleton*, Cambridge University Press, 1995 gives references for the masonry structure.]

Contents

Chapter 1

The Civil Engineer

The first modern civil engineer in Great Britain was John Smeaton (1724–1792); his life work was memorialised in 1994 by the dedication of a plaque in the north aisle of Westminster Abbey. Windows above commemorate Sir Benjamin Baker (the Forth Bridge), Parsons (the steam turbine), Lord Kelvin and Sir Henry Royce; nearby are memorials to Thomas Telford, James Watt, Isambard Kingdom Brunel and George and Robert Stephenson; this is the 'Engineers Corner' of the Abbey. Close at hand are architects (Sir George Gilbert Scott, Sir Charles Barry, John Pearson) and scientists and mathematicians – the monument to Newton is justly magnificent.

It is right that architects, engineers, scientists and mathematicians should be grouped together. Some activities involve all four of the professions, and, in particular, it is sometimes hard to distinguish between the work of engineers and scientists. Engineers use Newton's mathematics and Faraday's physical laws; engineers and scientists have a common technical language to describe the tools at their disposal. However, there is a difference in the way those tools are deployed. Scientists use the tools to deepen understanding of their own subject, while engineers use the same tools in order to do something, whether it be to design a turbine blade, an electronic circuit, or a radio telescope; to drive a tunnel under the Channel; or to create a great building – a Gothic cathedral or a steel-framed skyscraper.

Smeaton's scientific work, recognised by his election to the Royal Society at the early age of 28, resulted directly from his need to establish a theoretical basis for his engineering projects – it is in this sense that he may be described as a modern engineer. In the seventeenth century the Royal

Society, and in France the *Académie*, were in no doubt that the 'science' their members studied should be of immediate and practical use; this, indeed, was the whole intent of Francis Bacon's 'new philosophy'. However, the split between 'science' and 'engineering' widened quickly, and as early as 1783, for example, Cambridge University created a Professorship of Natural Experimental Philosophy to ensure that engineering developed as a discipline in its own right. Earlier, in 1749, a technical university had been established at Mézières in France in order to teach hydraulics, earthworks and surveying to young army officers (and the great *Écoles*, and the *Polytechnique*, were established before the turn of the century).

However, the theory taught in the universities in the eighteenth century was not adequate for those engineers engaged in 'the art of directing the great sources of power in Nature for the use and convenience of man' (to use Thomas Telford's words of 1828 as first President of the Institution of Civil Engineers). An engineer in charge of a major project, well-schooled though he may have been, was forced to make his own experiments and to develop his own scientific theory. Smeaton was particularly well-read, and his library included major works from both sides of the Channel (Newton's *Principia*, of course, but also Bélidor's *Architecture hydraulique* of 1735, Desagulier's *Experimental Philosophy* of 1744, a Vitruvius, and so on). However, it was known that a great range of engineering problems required urgent attention, and the Royal Society, and the *Académie*, offered prizes from time to time for their solution.

The four great problems in structural engineering throughout the eighteenth century were the strength of beams, the strength of columns, the thrust of arches and the thrust of soil (that is, the behaviour of soil behind a retaining wall, a problem in the field of what is now called soil mechanics). When Charles Coulomb, as a recent graduate of Mézières, was sent as a young army officer to fortify the island of Martinique (against the attacks, among others, of the British), he found that he lacked the theory for each of these four problems. He needed solutions in order to design his fortifications; on his return to Paris after nine years abroad, he presented his contributions to theory to the *Académie* in a notable paper of 1773. (The fourth section of this paper is a fundamental contribution to the science of soil mechanics, of which Coulomb is regarded by engineers as the founder. He is remembered by physicists – who do not know of him as a civil engineer – for his later work on electric charges.)

Smeaton, like Coulomb, was interested in a broad range of topics, and, throughout his life, he presented his scientific work to the Royal Society. Eighteen papers were published in the *Philosophical Transactions* between 1750 and 1788 (Coulomb read 32 memoirs to the *Académie/Institut* between 1773 and 1806). In his papers, Smeaton was concerned with three main topics: instruments, astronomy and mechanics. He was interested in problems concerned with navigation on the one hand, and with astronomical observation and instrumentation on the other, and his papers on these subjects, early and late in his career, were careful and intricate. Between these publications, however, three papers are of a different class; they deal, in essence, with fundamental questions of theoretical mechanics.

It was the great paper of 1759 (soon to be translated and published in France), 'An experimental enquiry concerning the natural powers of water and wind to turn mills', that was rewarded with the Copley Medal, the highest award bestowed by the Royal Society for original research – Smeaton was then aged 35. Coulomb's work had been in solid mechanics; Smeaton was dealing with fluids, and once again the basic science had not been established which could be used for engineering design. Smeaton knew of existing French theory, and his own experiments demonstrated clearly where this was incorrect. In fact his own theoretical enquiries at this time did not resolve some of the basic issues, and it was Borda in France, 10 years later, who contributed the mathematics which led finally to the turbine. What was hazy in 1759 were the concepts of momentum, energy and work, and how these should be expressed mathematically; Smeaton got far enough in his analyses to be able to make correct design decisions for mills.

In all of this scientific work are reflected the difficulties that confront an engineer when he expands a basic store of science, great though that may be, to tackle problems not met with before. Smeaton's eighteen papers are valuable in themselves, and they show how he was contributing to his profession, establishing ideas which could be understood, taught and used by his successors. But the papers are in effect by-products of the design and execution of a great range of civil engineering works.

Perhaps the best known of Smeaton's works is his first lighthouse of 1756/9, the Eddystone. His own *Narrative of the Building and a Description of the Construction of the Edystone Lighthouse*, published as a book in 1791, shows how an engineer achieves something new. The Trustees had to agree to a house built in stone; Smeaton himself went to Plymouth to survey the Rock (he invented his own surveying apparatus for this purpose);

Pl. I.

Sav. Etrang 1773 Pag. 382. Pl. XV.

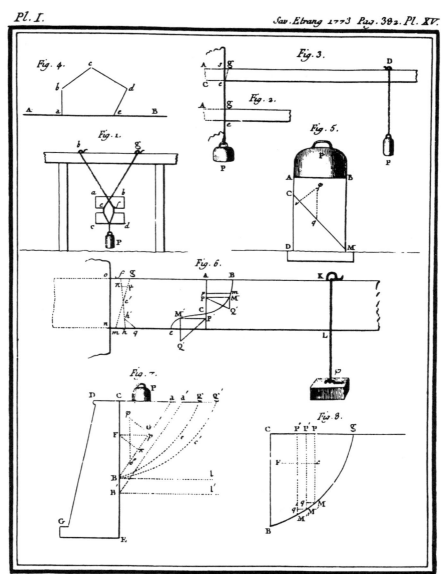

Plate I of Charles Coulomb's first scientific paper, presented to the French Academy in 1773 and published in 1776. The plate illustrates three of the four problems discussed by Coulomb – illustrations of masonry arches were grouped on a second sheet.

Figure 3 shows a beam, embedded in a wall at the left-hand end, and carrying a load near its tip – how could the breaking load of this cantilever beam be calculated? A tension test, shown in Fig. 1, would give the fracture strength of the material – could this fracture strength be correlated with the bending strength of the beam? Coulomb showed how such a prediction must be made, and Fig. 6 gives the technical illustration from which he developed his theory.

Similarly, Coulomb was concerned with the calculation of the breaking load of a stone column loaded vertically, as in Fig. 5, and he showed that fracture would occur along an inclined plane CM, at an angle which could be predicted from the properties of the material.

The third problem tackled by Coulomb was the design of retaining walls to hold back soil (Fig. 7). To design the wall, it was necessary to evaluate the thrust of the soil, and Coulomb's 1773 paper is regarded as fundamental to the development of the science of soil mechanics.

These four problems – the strength of beams, the strength of columns, and the thrust of soil, together with the thrust of arches – were the most important problems of civil engineering in the eighteenth century. They arose in a military context, in the design of fortifications, but their solutions opened up a range of applications in the general field of structural engineering.

SECTION *of the* EDYSTONE LIGHTHOUSE *upon the* E.&W.*Line, as relative to* N.&S.
on Supposition of its being LOW WATER *of a* SPRING TIDE.

Engraved in the Year 1763 by Mr. Edw.d Rooker.

Eddystone lighthouse: An engraving from a drawing by Smeaton.

The Eddystone Rocks are part of a reef of red granite, 14 miles south of Plymouth; they are largely covered at high tide, and have always been a hazard to shipping. A first lighthouse was built on the rocks in 1696, and was destroyed in a storm in 1703. A second lighthouse, in timber, was completed in 1709, and survived until 1755 when it was destroyed by fire. Smeaton was relatively unknown when, at the age of 31, he was commissioned to design a new lighthouse, and he was determined to use stone, both for its weight which would resist the forces of wind and sea, and because it would not burn.

The first problem Smeaton faced was to anchor the stones securely to the rock base. Six steps were cut in this rock, and the steps filled with massive blocks, weighing between 1 and 2 tons, dovetailed together and to the rock. The seventh course was the first complete course of the lighthouse; the exposed blocks were of granite, and the inner of Portland stone. Entrance to the house was above level 14, and a spiral stair led to the succession of four chambers above level 24; these living quarters were enclosed by a circular wall consisting of single blocks of granite. Access between chambers was by means of a movable ladder.

The lighthouse was commissioned in 1759, and survived for well over a century – it was then found that the base rock was in a serious condition. Smeaton's tower itself was in good condition, and the upper parts were taken down and rebuilt on Plymouth Hoe, after a replacement tower had been completed in 1882.

he devised an ingenious interlocking of each course of stone to confer stability on the whole structure; he made experiments, and finally determined the precise mix of lime and pozzolana to form a mortar that would set under sea water; and he considered the effects of wind and waves in his choice of a continuously curved profile for the lighthouse. Finally, he made sure that the work was prosecuted well and efficiently – engineers' tasks do not stop with the conceptual ideas, nor yet when those ideas have been clothed with design calculations and drawings; engineers must ensure that the projects are actually achieved.

For the twenty-five years following the completion of the Eddystone lighthouse, that is, from 1760 to 1783, Smeaton was intensely occupied with engineering projects. He designed more than 50 watermills and windmills, and some dozen steam engines for water supply and pumping. These are what would now be called mechanical engineering works, and of course gave rise to those questions of momentum and power to which he tried to find answers. In the civil engineering field, Smeaton designed four major public bridges, including the three fine masonry arches at Coldstream, Perth and Banff; there were also half a dozen minor bridges in stone or brick, and two aqueducts. He was responsible for the Forth and Clyde Canal, and for substantial improvements to several river navigations; and his works include major harbours, piers and fen drainage schemes. Before 1760 the profession of civil engineering hardly existed; a decade later recognisably modern consulting engineers could be seen, of whom Smeaton was pre-eminent. Such men, then as now, travelled to where their skills were needed; they made designs on the basis of their store of knowledge, and, if that store were insufficient, they made their own theory and experiments and contributed to the science of their subject; and they made sure, if necessary with assistants, that the work was properly done.

Engineers cannot work alone. Mention has been made of the Institution of Civil Engineers which in 1828 formally established a channel for the exchange of professional information. Much earlier, in 1771, an informal Society of Civil Engineers had been established; Smeaton was a founder member, and he was a regular attender of meetings until his death, when the group was renamed the Smeatonian Society. It was in essence an eighteenth-century dining club, and as such it still exists today. There was, however, a vital function – the club provided the opportunity for a group of civil (that is, non-military) engineers to discuss their work.

Chapter 2

Pre 'Scientific' Theory

Ars sine scientia nihil est. It is clear that the word *scientia*, recorded in the minutes of the expertise held at Milan in 1400, does not carry the full modern meaning of a system of thought related to the physical universe. The word 'science' today carries the implication of a more or less standard sequence – observations have been made, a hypothesis constructed, experiments designed to prove or disprove the hypothesis, some mathematical analysis undertaken, and finally a theory established which can be used to explain past observations and predict new results. Newton's gravitational theory is of this kind. Observed planetary orbits could be explained by the assumption of the universal inverse-square law, and future motion could then be predicted exactly (or almost exactly – the explanation for tiny departures from Newtonian cosmology had to wait for Einstein).

Babylonian astronomy was certainly a science. It was developed between 1000 and 500 BC, and based on accurate observations of the day-by-day positions of the sun and moon (and of the planets); numerical difference tables were then constructed from which future positions of the heavenly bodies could be predicted. Further, once the motions of the moon and sun had been determined, it was not difficult to predict when the moon would be close to the ecliptic at opposition or conjunction – and hence tables of lunar and solar eclipses could be established.

The medieval *scientia* of building was of this sort. The word implied scholarship and learning – knowledge acquired by the practice and study of construction. This knowledge, or 'theory', was recorded, and could be used in the design of new construction; embodied in a set of rules for building, the theory noted the way in which successful designs had been made in the past. Similarly, the word *ars* used at Milan referred to the mason's practical knowledge. Thus *ars sine scientia nihil est* – practice is nothing without theory – implies that construction, by practical men well-versed in their trade, should proceed in accordance with recognised rules.

It seems hardly necessary to stress that construction *must* have proceeded in this way. Vast Gothic cathedrals are clearly feats of structural engineering – it is out of the question that they could have been designed as acts of faith by untrained builders. Such feats were not achieved without *scientia*, nor without a master of the work who had studied for many years.

Ezekiel

The rules by which buildings could be designed were, no doubt, often passed on by word of mouth. They could also be recorded by drawing, although little survives that is more than a few centuries old. Written rules, however, appear surprisingly early. Chapters 40, 41 and 42 of Ezekiel, for example, record over several pages the sizes of gateways, courts, vestibules, cells, pilasters and so on, for a great temple; part of a building manual of 600 BC seems to be associated with the books of the Old Testament. From Ezekiel 40: 3 and 5: 'I saw a man holding a cord of linen thread and a measuring-rod ... The length of the rod ... was six cubits, reckoning by the long cubit which was one cubit and a hand's breadth.'

The dimensions which are given in the manual are in cubits and palms. The Hebrew cubit, about 17.7 inches or 450 mm, the length of the forearm from the tip of the middle finger to the elbow, was divided into six palms; the 'royal' cubit – a cubit and a palm – was therefore about 20.7 inches or 525 mm, very close to the Greek standard of seven palms. What the master builder – the architect – was holding was the 'great measure', without which work could not proceed on an ancient (or medieval) building site. This particular measure was six cubits in length (somewhat over 3 m), marked with a sub division of palms and further subdivided into fingers; it could be used to establish the major dimensions of rooms as well as smaller individual dimensions, merely by using the numbers listed so diligently in the books of Ezekiel. When such numbers had been recorded, either in a manual or in a drawing, then they could be transferred to the site once the great measure had been physically constructed.

The great measure – the rod – was a convenient practical tool. The British measures, learned by school children in the UK even after World War II, were

(3 barley corns	=	1 inch)
12 inches	=	1 foot
3 feet	=	1 yard
$5\frac{1}{2}$ yards	=	1 rod, pole or perch
40 rods	=	1 furlong
8 furlongs	=	1 mile

Barley corns had perhaps disappeared as units by the time of World War I – and the seventeenth century subdivision of the inch into twelve lines (*lignes* in France), had been overtaken for scientific work by metric standards. As noted in Ezekiel, the rod could be used as a unit for establishing large leading dimensions (as, in England, 4 rods = 1 chain = 1 cricket pitch), and also subdivided for detailed work. The essential feature of the great measure on a building site was that it was part of the building. It was not an absolute 'yardstick'; if it were cut slightly smaller, then a slightly smaller building would result from the same building plan. The same arguments would apply to a sculptor creating a human figure. The figure can be made to any scale but, whatever size is chosen, the ratio of one part of the sculpture, say the head, to any other part, say the hand, will be the same. Once the dimension of a single component of the statue had been fixed – the foot, for example – all other components can be expressed in terms of that foot. The foot is the unit of measure.

The materials

For those thinking in a modern 'scientific' way, it seems hard to believe that numerical rules, such as those of Ezekiel, could give rise to satisfactory designs for large buildings. The rules were essentially rules of proportion, but without any absolute scale – perhaps reasonable for a work of art, but difficult to conceive as applying to the stone used in building a large temple. The fact is that Greek temples have survived from the time of Ezekiel (for example, the Heraion, Olympia, 640 BC), and the mere evidence of survival gives powerful support to the view that numerical rules for masonry are of the right kind. As will be seen, twentieth-century structural theory confirms this view.

Masonry was one of the two main materials available for building, from the earliest times up to little more than a century ago. The other material was wood, and there are great differences between the properties and behaviour of timber and stone. First, stone is used in small pieces – of a size to be handled conveniently by one or two workmen – from which are constructed large, indeed very large, buildings. An exception may be found in the huge Greek architraves that span between stone columns; the Greeks did not use the arch, whose action is discussed in Chapter 3. In general, however, the large feats of structural engineering were achieved by the assembly of small building blocks.

By contrast, timber is available in long lengths, and those lengths have properties which reflect their organic origin – a tree, rooted firmly in the ground, must resist vertical gravity forces and high transverse loading due to wind. Thus long baulks of wood, whether they come from cantilevered branches or from the upright trunk, are capable of sustaining bending, which implies good tensile as well as compressive qualities. A tree trunk, thrown across a ditch, will serve as a crude bridge; poles may be lashed together to form a stone-age hut; flat roofs and floors may be constructed; and, more sophisticatedly, pitched roofs may be designed by connecting three timbers in a triangle – two rafters and a tie. The analysis of the timber structure moves the science into modern times, and will be considered further in Chapter 4.

Masonry does not possess the tensile strength of timber. Individual blocks of stone are indeed strong in tension, but the blocks are assembled into a coherent structural form either with no mortar, or with mortar which has little strength. Thus the masonry structure is well able to resist compressive forces – gravity loads, for example, can be passed on by one stone pressing on the next – but any attempt to apply tension will result in cracking of the joints or complete separation of the fabric. As an example, a masonry pier cannot be imagined to be hung from the top, clear of the ground; but the four actual crossing piers in a cathedral can share the burden of carrying the 10 000 tons of tower above.

It turns out that, as a matter of fact, the compressive stresses that arise in a major cathedral or a large-span masonry bridge are very low – low, that is, compared with the potential of the material. An illustration of the magnitude of stress to be expected may be gained from an examination of a parameter for strength used by nineteenth century engineers. Instead of the modern idea of a crushing force per unit area of the material, that is, a crushing stress, the parameter was expressed as the greatest height to which a prismatic column might, in theory, be built before crushing at its base due to its own weight. For a medium sandstone, this height is say 2 km; a granite pier might have a height of 10 km. The tallest Gothic cathedral has a height of about 50 m from the floor to the high stone vault – of course, a pier in such a cathedral carries the weight of that vault, of the timber roof above, wind loads and so on, in addition to its own weight. Even so, the 'factor of safety' seen through modern eyes seems enormous. In summary, and in very round figures, the four crossing piers in a cathedral supporting a massive tower will be working at an average stress of less than one-tenth of the crushing strength of the material. The main portions of the load-bearing masonry structure (say flying buttresses or the panels of the high vault) will be working at one-hundredth of the crushing stress, and infill panels and walls which carry little more than their own weight will be subject to a 'background' stress as low as one-thousandth of the potential strength.

A background of low compressive stress is essential for the stability of the masonry structure. The small pieces of stone are compacted by gravity into an overall shape determined by the architect, but that shape can only be maintained if the stones do not slip one on another. The elements may interpenetrate to some extent, or may be cut with some joggles or keys; in the absence of such designed roughness, the main element of stability is the low compressive stress which will allow friction forces to develop, locking the stones against slip. (It is evident that the stones themselves must have a certain minimum size; a dry stone wall can stand, but an attempt to build the wall from sand would be unsuccessful – the sand would slump away.)

Thus the essential structural features of masonry as a material are that it may be considered to have little or no tensile strength; that, on the other hand, it is subject to such low compressive stresses that the question of crushing strength does not arise in practice; and that there is sufficient internal friction for the shape of the structure to be maintained. These assumptions are, by and large, obeyed by real masonry constructions; exceptions may of course be found, and the theory must finally be broadened to include these exceptions. The main features, however, have important implications for the choice of particular stones as building materials. For example, tufa (a porous lightweight calcium carbonate) and clunch (a soft easily-worked limestone, as chalk) may be used appropriately despite their poor strengths; a modern example is the use of breeze-blocks. Similarly, sun-dried mud (adobe), with or without straw reinforcement, may be used to construct 'high-rise' buildings, as in the Yemen.

As an example of a different technology, non-reinforced concrete (mass concrete) may also be regarded as masonry, not as being composed of small elements, but cast into a continuous structure which hardens chemically into an 'artificial stone'. The use of concrete is by no means new; the Romans used a mixture of lime and pozzolana (a naturally occurring volcanic earth) to produce a good mortar for binding together squared or rubble stone, or gravel – a mortar, moreover, which is resistant to the effect of sea water (which usual modern cements are not – as was noted in Chapter 1, Smeaton devoted considerable effort to designing a suitable mortar for his Eddystone lighthouse). Although mass concrete seems to be continuous, to be, in fact, monolithic, the material remains weak in tension. Thus, to anticipate an example to be discussed in Chapter 3, a masonry dome may be made of brick, as at Florence, or of stone, as in St Peter's, Rome, or it may be made of mass concrete, as in the Roman Pantheon (120 AD). All of these domes, including the Pantheon, exhibit significant cracking where the structures have attempted to develop tensile stresses – the monolithic Pantheon has been broken into smaller pieces, but still retains its structural integrity (as do the other two domes).

This, then, is the material for which Ezekiel records rules of construction. Such rules, this *scientia* of building, may be traced in recognisable form throughout the next 2000 years.

Vitruvius

Vitruvius (c. 30 BC), writing five centuries after Ezekiel, makes *ordinatio* the first of his six principles of the theory of architecture, and it becomes clear that *ordinatio* is nothing other than the great measure, the rod made up of modules (*quantitas* in Vitruvius) by which the whole building may be measured and constructed. He then develops his theory of proportions in terms of modules taken from the great measure.

For example, if the diameter of a column in an araeostyle temple has a diameter of one unit, then the intercolumniation, the distance between the two central columns, should be four units, and the height of the columns should be eight units. Further, for a column of the Ionic order, the height of both the base and the capital should be half a unit, and there are further and more subtle subdivisions for finer details of the work. Doorways are treated in similar ways, with refined rules for almost imperceptible taper from the ground to the lintel. Vitruvius may not have been himself a great architect, but his compilation of earlier Greek writings, moulded by his own experience, became in the following centuries the standard text on building. Echoing Ezekiel, he showed how the ground plan of a new construction could be laid out by use of the great measure; how the proportions of every part of the plan and elevation could be determined by the modules scribed on the measure; and how the aesthetics of the design should form part of the whole process. Finally, nothing can be built without practical experience, and in his first chapter, Vitruvius stresses that *fabrica* (practice) and *ratiocinatio* (theory) are both necessary for the training of an architect. Indeed, after stating his principles and before embarking on his theory of proportions, Vitruvius discusses in detail the various materials available, including brick, sand, lime, pozzolana, stone and concrete.

Vitruvius himself had seen service as a military engineer with Julius Caesar, and was skilled in the construction of military engines – he mentions ballistae, scorpiones and other artillery. He considers that architecture is properly concerned with three matters: the construction of buildings, the making of time-pieces and the design of machinery. His manual was read throughout the medieval period, and was copied again and again for use in monastic schools and, above all, in the masonic lodges. The hand of Vitruvius is clearly visible in the fragments of Gothic lodge books that have survived.

Villard

The thirty-three remaining leaves of the sketchbook of Villard de Honnecourt, compiled about 1235 and later, deal precisely with buildings, clocks and machines. The book was written for skilled professionals, and is infuriating for modern readers – it assumes that the reader is already familiar with the basis of Gothic design, and is therefore silent on matters of fundamental interest.

Villard, like Vitruvius, was a minor architect; their manuscripts have survived, but nothing is known of any of their building work. What is clear, however, is the thread that ties the two together. Villard sketches timber roofs, shows examples of the 'engines' used for his trade – for example,

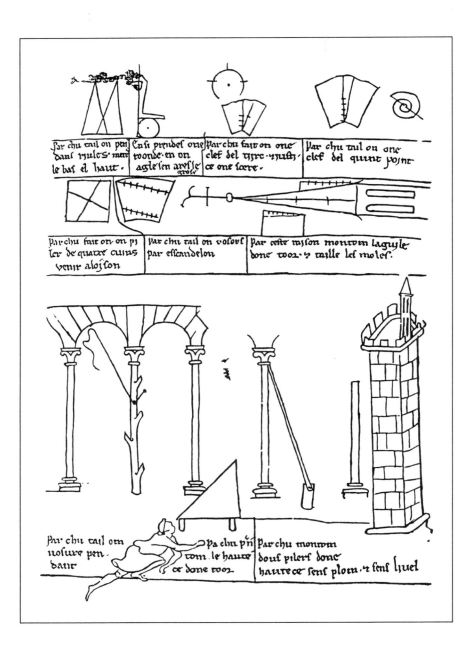

Page 40 of the sketchbook of Villard de Honnecourt, 1235 and later.

The centre caption at the bottom refers to the survey under way: Pa chu p'ntom le hautece done toor – *Par ce moyen on prend la hauteur d'une tour* – how to take the height of a tower. The standards of lining, levelling and plumbing employed in the construction of Gothic cathedrals were outstanding. On the right of the figure an arcade is being set out – how to set up two piers without plumbline or level. On the left is one of those medieval constructions designed to astonish and delight: Par chu tail om vosure pendant – *Par ce moyen on taille une voussure pendante* – how to construct a hanging voussoir; the final removal of the tree trunk leaves a capital suspended magically in mid-air.

The small sketches above all deal with problems of mensuration, of which the easiest concerns the spire. The example sketched shows a spire of height four times its base – then the slope of each side is one in eight. The sketch at top right concerns an irrational dimension which cannot be found on the great measure: Par chu tail on one clef del quint point – *Par ce moyen on taille une clé de quint-point* – How to cut the key-stone for a fifth-point arch.

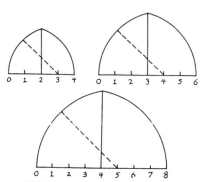

Third-, fourth-, and fifth-point arches, numbered in the presumed thirteenth-century manner.

The height of a fifth-point arch, and (half) the angles of the faces of the keystone, contain the irrational square root of six. The required dimension (and also for third, fourth … point arches) can be measured from the sketched spiral – this is not a true spiral, but is made up of successive semicircles drawn from two centres one unit apart. Frequently used "irrational" angles could be incorporated in the mason's square, such as the one shown in the top row of sketches.

power-driven saws – and he exposes geometrical rules of construction (as does Vitruvius). A large number of pages are devoted to portraiture, with Gothic faces and intensely sculptural drafting of drapery. The lions that appear in these sketches are fanciful – 'drawn from life' says Villard, although it is obvious that no real lion had actually served as a model. Birds, dogs, horses and ostriches abound. Villard may not have seen all these animals, but he had journeyed – to Hungary, for instance, a great voyage for a country lad who had been apprenticed in Picardy – and on these journeys he recorded, for his lodge, the new and extraordinary inventions of the 'golden age' of Gothic, which lasted for a century and a half, from 1140 (the abbey church of St Denis) to 1284 (the collapse of Beauvais).

A key problem exposed in Villard's manuscript concerns a geometrical construction for the doubling (or halving) of a square. Once again, this is precisely a question discussed by Vitruvius, who ascribes the solution to Plato. A piece of land ten feet long and ten feet wide, says Vitruvius, has an area of one hundred square feet; what is the length of the side of a square which has area two hundred square feet? The answer, of course, is the diagonal of the first square, of length $10\sqrt{2}$ feet. Illustrations of this are found in Villard's sketch book.

The problem seems trivial, but it exposes a fundamental mathematical difficulty. A great measure, divided into cubits, subdivided into palms, and further divided as finely as one pleases, should (it might be thought) provide lengths which can be transferred on the building site to elements of any conceivable dimension. This is not so, and the fact was recognised by Greek mathematicians. The irrational numbers, of which the square root of 2 is one example from an infinity of such numbers, cannot be so measured – there is no mark that can be made on the great measure to represent $\sqrt{2}$, but the commentaries in Vitruvius and in Villard show how such a length can be constructed. The matter, viewed through present-day eyes, is of no practical consequence, but it is intellectually fascinating, and Pythagoras's beautiful and simple proof of the irrationality of $\sqrt{2}$ has survived over two thousand years. For medieval builders, however, it was a problem to be resolved.

Milan Cathedral

Many similar problems of mensuration are dealt with in Villard's sketch-book, and it was problems of mensuration that became acute in the building of Milan Cathedral. The cathedral was started in 1386, over a hundred years after the end of the High Gothic period, and difficulties of construction led to two well-documented expertises, in 1392 and 1400. The original design was *ad quadratum*; that is, the height of the work to the top of the high vault should be the same as the total width of the nave and four aisles – the transverse section was to be contained within a square. In 1391 the work had advanced to the point where the height of the piers had to be finalised, and doubts were expressed about the original intention. The Milan lodge sought advice from that of Cologne, but finally accepted recommendations from Stornaloco, a mathematician from Piacenza.

The clear (internal) width of Milan Cathedral is 96 braccia (the braccio, or the arm, the Milanese cubit, is just under 2 feet, or about 0.6 m). Stornaloco proposed that the height should be 84 braccia, that is, that the construction should be *ad triangulum*, with the cross-section contained within an approximate equilateral triangle. It was with this question of approximation that the Milan lodge needed advice from a mathematician. A true equilateral triangle on a base of 96 braccia has an irrational height, not measurable by the great measure, of approximately 83.1 braccia – Stornaloco recommended 84 braccia instead.

In summary, the architect's great measure for the grand plan at Milan was a rod 8 braccia in length, and the designation *ad quadratum* means that the same great measure of 8 braccia was originally intended for the elevation. Stornaloco's proposal of 84 braccia not only eliminated the irrational square root of 3 but also, in effect, fixed a great measure of 7 braccia for the elevation. The proposal was disputed, and some wished to revert to the *ad quadratum* form; the Italian experts, on the other hand, wished to reduce the great measure for the elevation still further, to 6 braccia. In the event, they accepted Stornaloco's figure of 28 braccia for the height of the piers of the outer aisles (reduced in the event to $27\frac{1}{2}$ braccia to accord more exactly with the equilateral value of about 27.7), but above this level the work was completed to a vertical great measure of 6 braccia. As the horizontal great measure was unalterable at 8 braccia, the mensuration above the level of the piers was 'Pythagorean'.

All of this expertise was concerned, then, with establishing the *ordinatio* for the work, so that the stones could be cut to fit exactly into the determined overall dimensions, and so that the dimensions of each component of the work could be laid out by reference to the great measures. The whole theory of building lay in the numerical rules of proportion determined by

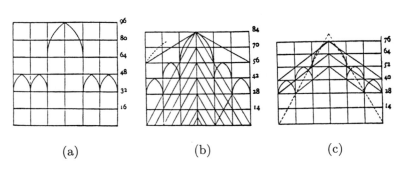

(a)   (b)   (c)

Milan Cathedral

The cathedral was started under the patronage of the Duke of Milan in 1386; despite the Visconti money, building was slow, and it was not for 5 years that a final decision about the height of the building had to be taken. The great measure, the actual wooden rod used to establish major and minor dimensions, was 6 braccia long – the braccio was about 0.6 m, so that the great measure was 4.8 m, very close to the English rod, pole or perch of $16\frac{1}{2}$ feet. The width of the cathedral had been set out as 12 rods, or 96 braccia, and the original intention had been to construct *ad quadratum* to the same height of 96 braccia, shown in Fig.(a). In 1392 Stornaloco, a mathematician, recommended that construction should be *ad triangulum*, with the cross-section that of an equilateral triangle of height $48\sqrt{3}$ or approximately 83.14 braccia, an irrational number which could not be set out by the great measure. Stornaloco proposed that the height should be 12 rods of 7 braccia, or 84 braccia, as in Fig. (b) – a new great measure was constructed for vertical as opposed to horizontal measurement. In the event, this measure was used up to a height of 28 braccia; the great measure was then reduced to 6 braccia, and the cathedral was completed as in Fig. (c).

the measures. Of great interest is the fact that one *ordinatio*, of 8 braccia, was used for horizontal dimensions, and another *ordinatio*, of 7 braccia for the lower work and 6 for the upper, for vertical dimensions.

Building proceeded reasonably smoothly until 1399 when again a major dispute led to the holding of another expertise. On this occasion, Giovanni Mignoto (actually Mignot) came from Paris, and Giacomo Cova from Bruges; they were joined within the next year by eight Italian architects to form a full-scale commission of enquiry. Mignot started, at the turn of the year 1399–1400, by drawing up a list of 54 points in which he found the work at Milan to be defective. The second half of the list consists essentially of trivia, but even the first half, of greater importance, is curiously arranged. Mixed up with objections which, if correct, are clearly serious – that the buttressing was insufficient, for example – are objections of a different kind – that the canopies were set too high above the carved figures, or that the capitals and bases of the piers were not in the right proportions. These points are given equal prominence and are dealt with equally seriously by the Italian defenders of the work. Mignot was not satisfied with the responses, and it does seem that the Italians were inventing arguments to support their views, rather than appealing to some more rational and absolute base for discussion.

The Italian assertion ... *archi spiguti non dant impulzam contrafortibus* particularly upset Mignot as a response to his criticism that the buttressing was weak. As a reply to this criticism, the first defence was that the counterforts, the main buttresses, were well-built of strong stone, connected with iron cramps, and well-founded, so that they were indeed strong enough. Secondly, without prejudice to this defence, the buttresses were really not necessary, since pointed arches do not thrust (*archi spiguti* ...). Finally, it had already been decided to tie the heads of the columns with strong iron ties to absorb any possible thrust (these and later ties are prominent in the cathedral today).

There are grounds for saying that the thrust of a pointed arch is less than that of a round arch of the same span, but Mignot may perhaps be excused his bad temper. He certainly emerges as a much deeper scholar, versed in the theory of building, to the point where the Italians sulkily fell back on the statement *scientia est unum et ars est aliud* – theory is one thing and practice another – Mignot's rules were all very fine, but they actually knew how, in practice, to build a cathedral.

Mignot's reply, *ars sine scientia nihil est* – practice is nothing without theory – seems to herald the dawn of a new age of architecture. It was, in fact, nothing of the sort – it was a final statement based on the experience of two millennia of structural engineering. Mignot had a book of rules with him which governed the design of great churches – his lodge book; this was his *scientia*. He had applied those rules to the work as he found it at Milan, and by those rules he had found it wanting. He may have had a more comprehensive set of rules, or possibly better rules, than the Italians, but it seems clear from the records of the expertise that Mignot himself did not understand his own scholarship. His mixing together of 'aesthetic' and 'structural' criticisms implies that he did not, in any deep sense, understand either sort of rule, or even distinguish between them – he merely knew that rules were being broken. Mignot's rule book had probably been compiled one or two centuries earlier, in the middle of the High Gothic period, and had lain in the lodge as a dead manual whose purpose was increasingly dimly perceived.

The Milan lodge rejected external advice on major problems of construction, and continued to build by their own rules. The cathedral has survived for six centuries.

The Renaissance

Structural engineering was design by numbers – the fifteenth century saw the end of this unbroken tradition of 2000 years, from Ezekiel (and earlier) through Vitruvius to the secret books of the masonic lodges. It seemed, at first, that the tradition was merely being re-interpreted. Alberti's ten books *On the Art of Building* were completed in 1452 and published in 1486 – Alberti is not kind to Vitruvius, but in fact he reinforced Vitruvius's authority, and stressed, above all, the importance of proportion for correct and beautiful building. Brunelleschi had made exact measurements of classical buildings in Rome; now, with the invention of printing, it became possible to publish an illustrated Vitruvius, and the Renaissance of Roman architecture was under way. The Gothic rules were complicated – as John Harvey has said, ' ... so complicated that no one who had not served a long apprenticeship and spent years in practice could master them; whereas the rules of Vitruvius were so easy to grasp that even bishops could understand them, and princes could try their hand at design on their own.'

With the rules in one hand, and illustrations in the other, it was possible for an educated man to become a successful architect, without the trouble of actually learning how to build. Considerations of history and aesthetics started to become divorced from engineering structure, in a way that would have been unthinkable to the medieval 'master', who knew in the fullest technical sense how to handle his material, as well as how to give the building an 'architectural' design. It is from the Renaissance that the professions of the architect and engineer begin to diverge – both spring from Gothic, but the architect concentrates on the rules of proportion embedded in the theory, while the engineer begins to explore the scientific rules embedded in the practice of building.

Printing also furnished another nail in the Gothic coffin with the wide dissemination of the secrets of the lodges. Roriczer's (1486) *The Right Way for Pinnacles (Der Fialen Gerechtigkeit)*, for example, is a textbook for apprentices written by the cathedral architect at Regensburg. The exercise given is simple, but it reveals permanently those rules for the construction of irrational lengths and for subdivision of the basic module that were at the heart of the apprentice's training. But the rules were, in any case, no longer needed – modern science had started, and the decimal notation had been learned from the Arabs. The irrational value of the square root of 2 could now be measured to any degree of practical accuracy on a decimally subdivided rule.

The lodges did not take kindly to science. They resisted the 'infiltration' of new ideas – Wagner's *Die Meistersinger* shows how a medieval guild might have reacted, and indeed Mignot at Milan seems a sort of Beckmesser, insisting on rules being obeyed to the letter. The 'Gothic' carpenter in Britain, well into the second half of the twentieth century when the country adopted the metric system, was still measuring work in eighths of an inch, or, if greater accuracy were needed, in sixteenths – the basic module was subdivided in the traditional way. The masonic lodges, like the British carpenter, were content to copy their rules from generation to generation, forgetting, like Mignot, the genesis of those rules.

Chapter 3

Arch Bridges, Domes and Vaults

Ancient building works included roads, fortifications and harbours; even on a small scale, the builder needed a good knowledge of the properties and use of appropriate materials. At a larger scale, when the works may be thought of as civil engineering rather than building, rules were needed as well as practical knowledge. The design of a cathedral required both *scientia* and *ars*. In the same way, a tree-trunk thrown across a small stream will provide a serviceable bridge, but bridges of larger span must be designed and require the use of specialised theory.

Perhaps surprisingly, two sophisticated types of bridge have been in use for at least 6000 years. In the suspension bridge, a 'deck' is hung from 'cables', and the form has hardly changed through the millennia, although the materials are no longer vegetable. Originally, cables were made from creepers or lianas, or from ropes of twisted strands, made in turn from fibrous plants, and the walkway (if it existed at all) was made from wood. The structural action of such a bridge is straightforward and, so to speak, visible in the form of the bridge; the loads on the walkway are transferred to the cables by hangers, and the cables act in tension, finally pulling on their anchorages at either end of the span.

The other early type of bridge is the arch. Originally, perhaps in Mesopotamia in 4000 BC, the material used was brick, although the bricks were sun-dried rather than baked. Similar use of bricks was found in Egypt a few hundred years later. Burnt bricks followed, for both arches and vaults, and the Egyptians first used cut stone about 3000 BC. Much later, the Etruscans had mastered the art of cutting wedge-shaped voussoirs, and assembling them in arches and vaults. All of this technology was absorbed by the Romans in the expansion of their empire, and by about 500 BC they were constructing large-span masonry bridges.

Vitruvius gives no theory for the design of such arches, but rules, and probably numerical rules, must have existed. It is likely that Vitruvius had not himself been concerned with the building of large arches, such as those used, for example, in aqueducts – he does, however, discuss arches on a domestic scale. The problem is to make an opening – for a door or window – in a flat stone wall; how is the wall to be carried over such an opening? 'We must discharge the load of the walls by means of archings composed of voussoirs with joints radiating to the centre,' said Vitruvius. Almost all Roman arches were semicircular, and the joints between voussoirs all radiated from the centre of the circle. The term 'centering', used for the constructional process, came to be applied to the falsework necessary to sustain the arch while it was being built. On the domestic scale, the completed arch formed the head of the doorway or window.

The practice and theory of Roman bridge building came to be vested in a religious institution, the *Collegium Pontifices*, which eventually controlled roads as well as bridges. The head of this institution was the *Pontifex Maximus*, still one of the titles of the Pope. Although the *Fratres Pontifices* would not have calculated forces, they would have been well aware that the arch behaves in a way exactly opposite to a suspension bridge – the arch thrusts at its abutments rather than pulling on anchorages. One of the major problems in arch bridge design is the construction of the abutments to resist the arch thrust and, for a multi-span bridge, the construction of internal piers in the river bed.

The bridge-builders' secrets, just as the masons' secrets, survived the dark period of the Middle Ages; the *Fratres Pontifices* surfaced, in the twelfth century, as the Benedictine *Frères Pontifes*, and an order was soon established in England. Peter, Chaplain of St Mary's, Colechurch, started to build the first stone London Bridge in 1176, replacing earlier timber bridges – the 19 pointed arches survived until the nineteenth century. The problems of Old London Bridge were mainly with the river piers and their enclosing timber 'starlings' – the piers blocked the river and caused fast flow through the arches, scouring the foundations. The masonry arches themselves, with their superload of shops and dwellings, seem to have been satisfactory.

An illustration from a twentieth-century manual on masonry, showing how to set out a semicircular arch. The arch consists of seven stones, shaped to key with each other and to suit the horizontal courses of the ashlar wall. Temporary formwork – the centering – must be used to support the stones until the central keystone is placed. The joints between the seven stones radiate from a common centre, following a practice of 2000 years and as recommended by Vitruvius's building manual of about 30 BC. Gothic pointed arches do not have a common centre, and the profiles are not necessarily circular, but the term 'centering' still survives for the temporary falsework.

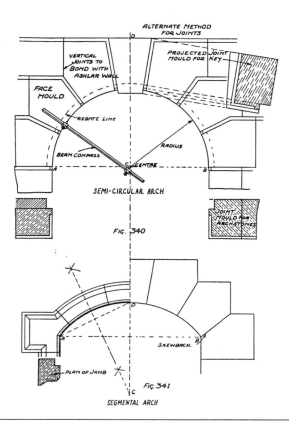

Masonry arches

It was with the estimation of arch thrust that the first numerical calculations were concerned, as well as with the attempt to understand arch behaviour in general. As will be seen, recognisably modern work had been done before 1717, but it was in that year that Gautier's book on bridge abutments was published; he summarised the 'five difficulties' that needed resolution:

1. The thickness of abutment piers.
2. The dimensions of internal piers as a proportion of the span of the arches.
3. The thickness of the arch ring.
4. The shape of the arches.
5. The dimensions of retaining walls to hold back soil.

Gautier's first problem requires for its resolution a knowledge of the value of the abutment thrust from the arch, and his problems 3 and 4 recognise that the solution depends both on the shape and the thickness of the arch. Problem 5 widens the enquiry to a field of civil engineering which is now treated by the sciences of geotechnical engineering and soil mechanics.

Robert Hooke, in 1675, addressed the structural problems of the arch. He published in that year a Latin anagram concerning the 'true ... form of all manner of arches for building, with the true butment necessary to each of them'. The anagram gives *Ut pendet continuum flexile, sic stabit contiguum rigidum inversum* – 'as hangs the flexible line, so but inverted will stand the rigid arch'. Publication in anagram form was common in the seventeenth century. The scientific climate was competitive, and scientists were concerned to establish their priority in obtaining solutions to both old and new problems. However, it was important to guard closely any clues; a revealed hint might enable another worker, perhaps more adept mathematically, to reach the solution first. In the case of the arch, Hooke had a *new* idea, which he did not wish to give away – the mathematics of the statics of a hanging cord in tension are the same as that of the arch in compression. In other words, although Hooke did not state this, there is a fundamental correspondence between the suspension bridge and the masonry arch. If the mathematics of the 'catenary' could be solved, then at the same time not only would the shape of the arch be derived (Gautier's problem 4), but the all-important value of the abutment thrust could be determined (problem 1).

The mathematics of the hanging chain is not easy, and Hooke never accomplished the work, although it is clear that he had a complete physical understanding of the problem – indeed, in 1670, earlier than his published anagram, the minutes of the Royal Society record that Hooke had given an experimental demonstration of the principles of the arch. Contemporary mathematical giants (Leibniz, Huygens, John Bernoulli) did in fact obtain solutions, although they too were secretive about their work. A less perfect treatment was published openly by David Gregory in 1697 – his discussion is of the utmost importance. In Ware's (1809) translation from Gregory's Latin, he states:

'In a vertical plane, but in an inverted situation, the chain will preserve its figure without falling, and therefore will constitute a very thin arch, or fornix; that is, infinitely small rigid and polished spheres disposed in an inverted arch of a catenaria will form an arch; no part of which will be thrust outwards or inwards by other parts, but, the lowest part remaining firm, it will support itself by means of its figure. For since the situation of the points of the catenaria is the same, and the inclination of the parts to the horizon, whether [hanging] or in an inverted situation, so that the curve may be in a plane which is perpendicular to the horizon, it is plain it must keep its figure unchanged in one situation as in the other. And, on the contrary, none but the catenaria is the figure of a true legitimate arch, or fornix. *And when an arch of any other figure is supported, it is because in its thickness some catenaria is included.* Neither would it be sustained if it were very thin, and composed of slippery parts. From Corol. 5 it may be collected, by what force an arch, or buttress, presses a wall outwardly, to which it is applied; for this is the same with that part of the force sustaining the chain, which draws according to a horizontal direction. For the force, which in the chain draws inwards, in an arch equal to the chain drives outwards.'

Here, then, is Gregory's complete grasp of the end to which the analysis of the arch is directed; the horizontal component of the abutment thrust of an arch has the same value as the horizontal pull exerted by the equivalent hanging chain.

Further, the italicised sentence (italics added by Ware) is extremely powerful (and addresses Gautier's problems concerned with the shape of the arch and the thickness of the arch ring); Gregory asserts that if the shape of the inverted hanging chain lies within the masonry of the arch, then the arch will stand. This simple statement might seem, perhaps, to be little more than a reinforcement of medieval structural theory – the shape must be correct if the structure is to stand. Broadened, however, the statement furnishes the fundamental theorem of structural mechanics, which had to wait until the twentieth century for its formal mathematical proof – it is only necessary for the structural engineer to be assured that a structure *can* stand; if it *can*, then it *will*. This 'safe theorem' stems from the plastic theory to be discussed in Chapter 7, but its existence has perhaps always been evident to structural engineers, as it was to Gregory.

The inverted hanging chain is called the line of thrust of an arch; it represents the way in which the compressive forces are transmitted within the masonry to the abutments. When this line of thrust has been located for a given arch, then simple equations of statics enable the engineer to calculate the magnitude of the thrust of the arch, and the design may be completed. Work proceeded throughout the eighteenth century to try to locate the thrust line, and the clarifying concept of a 'hinge' was introduced.

The thrust of an arch on its abutments will cause those abutments to give way slightly. The only way the arch can accommodate the increased span – within the assumptions of a rigid material having no tensile strength and infinite compressive strength, and without slip of the stones on each other – is by cracking, that is, by the formation of effective hinges. The general shape of the line of thrust is known – it is that of the hanging chain. Since it must pass through the hinge points, it is uniquely located once those hinge points have been found.

However, the hinge points themselves are not uniquely located. If the span of the arch should decrease, as it might if it were part of a multi-span bridge, then a different arrangement of hinges would be produced in the arch, and the line of thrust will move to a new position. Each position of the thrust line corresponds to a different value of abutment thrust – if the

A simple masonry arch is made from identical wedge-shaped voussoirs – it is built on falsework, since it cannot stand until the last stone, the keystone, is in place. Once complete, the falsework (the centering) may be removed, and the arch at once starts to thrust at the river banks. Inevitably the abutments will give way slightly, and the arch will spread.

The lower figure, greatly exaggerated, shows how the arch accommodates itself to the increased span. The arch has cracked between voussoirs – there is no strength in these joints, and three hinges have formed. There is no suggestion that the arch is on the point of collapse – the three-hinge arch is a well-known and perfectly stable structure. On the contrary, the arch has merely responded in a sensible way to an attack from a hostile environment. In practice, the hinges may betray themselves by cracking of the mortar between the voussoirs, but larger open cracks may often be seen.

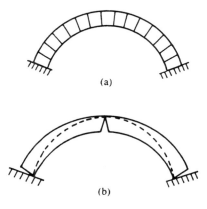

(a)

(b)

A model arch was demonstrated by W H Barlow at the Institution
of Civil Engineers in London in 1846. The model had six voussoirs,
and the 'mortar' in each joint was in the form of four small pieces
of wood, each of which could be removed by hand. Three out of
the four pieces were then indeed removed – since the forces must be
transmitted through the remaining piece, the line of thrust became
'visible'. Three different positions of the thrust line were sketched
by Barlow – the steepest curve, touching the crown of the arch at
the extrados, he called 'the line of resistance', and the flattest curve
'the line of compression'. These two lines represent the least and
greatest values of the horizontal component of the abutment thrust,
and correspond to slight movement apart or approach respectively of
the abutments.

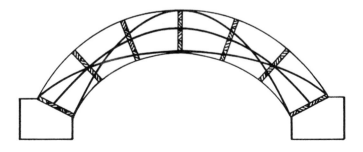

arch spreads, the thrust falls to a minimum value, and if the arch is forced to suffer a decrease in span, the thrust rises to a maximum.

The idea that structural quantities, such as the value of the abutment thrust, could have upper and lower limits, is not new; Coulomb, for example, made use of the concept in his 1773 examination of the arch. However, there was no discussion of the question as to which of the lines of thrust was 'correct'. It turns out that very small movements of the abutments of an arch can switch the thrust line from its 'maximum' to its 'minimum' position. This feature of the actual behaviour of a masonry arch was either ignored or not appreciated by the analysts.

The assumption of hinge points unlocked the statics of the problem, and enabled caculations to be made; very often, and certainly by the middle of the nineteenth century, a 'central' position of the thrust line was assumed, and this line was then 'clothed' with a suitable arch ring. Standard design tables enabled the engineer to construct elegant, economical and safe bridges, but by this time the masonry arch was already obsolescent. Iron Bridge had been built at Coalbrookdale in 1779, and Telford had projected a cast-iron span of 600 ft for the new London Bridge – in the event Rennie's 1831 masonry bridge replaced Old London Bridge, and has itself now been taken down.

More general techniques of structural analysis were also being developed in the nineteenth century – when applied to the arch, these purported, as will be seen, to determine the 'actual' position of the thrust line.

Domes

An arch bridge (or a barrel vault) is essentially a two-dimensional structure; the barrel can be cut, hypothetically, into a large number of parallel slices, each of which is identical to its neighbour. Once the structural action of a slice has been understood, then the structural action of the complete bridge or barrel vault is clear.

The simplest form of dome is generated by rotating an arch about its central axis – a semicircular arch will generate a hemispherical dome. The generating mathematical curve need not, of course, be a circle – a parabolic arch will give rise to a paraboidal dome, and a hen's egg has a more complex profile. However, whatever the profile, there are great differences between the behaviour of a two-dimensional arch and a three-dimensional dome, and these differences are reflected not only in the structural action but also in constructional procedures.

John Fitchen's conjectural falsework for the Pont du Gard.

The central span is some 25 metres. The centering rests on voussoirs which project inwards from the arch ring, and also has to be supported from below by scaffolding. The masonry arch cannot carry its own weight until all the voussoirs are in place – at this stage, however, because of the method of construction, the weight of the arch is actually carried by the timber centering. It is only when this is removed that the arch will begin to work structurally – the problems of de-centering while the full load is being carried are formidable. The gaps shown between the scaffolding and centering were filled with wedges, which could be knocked out to effect a controlled transfer of load. (Similar problems must be overcome in the launching of a newly completed ship.)

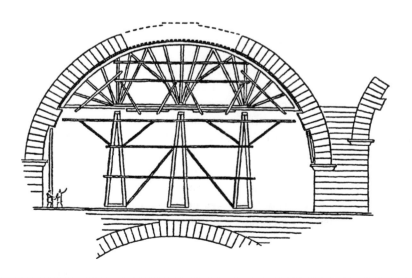

For example, the first few stones placed during the construction of a masonry arch may well be able to sustain themselves, each resting on the one below. As the arch curves inwards, however, a newly placed stone would be in danger of slipping, and the need for falsework, the centering, becomes clear. Any stone in the completed arch ring may be regarded as the 'key' stone, ensuring stability of the whole – but the keystone is normally thought of as the central stone in the ring, last to be placed and often given architectural prominence by its size and decorative carving.

Construction of a dome is easier. Although the profile curves inwards, a completed ring of stone not only rests on the finished courses below – it has no possibility of slipping. The complete ring is virtually incompressible, and any tendency to slip down is resisted by each stone being supported laterally by its neighbours on either side. Thus circumferential or 'hoop' forces can be generated along the lines of latitude of the dome, and these can exist in the presence of the compressive forces along the lines of longitude, the meridians (which are the counterparts of the compressive forces in the two-dimensional arch). Very little falsework is therefore needed during construction – once a ring is complete, any temporary supports can be removed, and re-erected ready for the next course to be built. There is no counterpart of the keystone of the arch; the dome can be terminated when it is incomplete, leaving an 'eye' open to the sky as in the Roman Pantheon, or in the innermost of Wren's triple domes for St Paul's in London.

The profound difference between an arch and a dome is illuminated by pursuing the idea of Hooke's hanging chain (as indeed Hooke himself pursued the idea). A flexible chain will take up a new shape to accommodate an applied load – a point load applied to a heavy chain will deform a 'catenary' into a more complex shape, and the line of thrust in the arch will shift correspondingly. By contrast, the three-dimensional analogous 'hanging membrane' is not deformable in this way – a sheet of cloth may be woven into the shape of a hemisphere, but that shape is then fixed, at least if the cloth does not wrinkle. That is, the cloth hemisphere may be suspended and partly filled, say with water – the shape will stay hemispherical, although the values of the forces in the threads in the warp and woof (the latitude lines and meridians) will depend on how much water is being carried. In mathematical terms, an arch or barrel vault is developable – the vault may be constructed by bending a flat sheet of paper (and even given a Gothic crease at a pointed crown). A dome is non-developable – it cannot be made from a flat sheet of paper without cutting and gluing and, once cut and glued, it is stiff.

Cracking in the dome of Santa Maria del Fiore, Florence

Brunelleschi's dome at Florence (c. 1420) is not a dome of revolution, but segmental — each of its eight segments, triangular in plan, is part of a barrel. There are eight groins where these barrels meet, which tend to 'collect' the weight of the dome, concentrating the forces at the corners of the supporting octagonal drum. The supports have given way slightly, and the dome has cracked to accommodate the increased span. The dome is buttressed to the north, south and east by domed apses, and on the west by the nave of the cathedral, and the crack pattern indicates that the north buttressing system has moved to the north, the east to the east, and the south to the south, with the west side perhaps remaining reasonably firm.

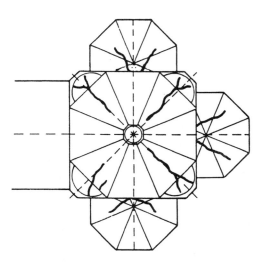

The wrinkles in a suspended membrane might develop if there were a tendency for the hoop forces to become compressive in certain regions under certain loads – the fabric is good at sustaining pulls but cannot tolerate compression. The corresponding behaviour in the dome is shown by the tendency for cracks to occur along the meridians, and this is exacerbated if the supports of the dome spread under the imposed loading. Stresses near the crown of a dome, whether it has an eye or not, are compressive in both the meridional and the hoop directions, but from an angle of about 25° down to the base, cracks may be seen in many actual domes.

For example, the Roman Pantheon is, effectively, of 'monolithic' concrete construction, and concrete, like brick or stone masonry, is weak in tension. Cracks are well-developed in the Pantheon. Similar meridional cracks in the dome of St Peter's, Rome, were the cause of concern in the mid-eighteenth century, but were analysed correctly in an extensive report by Poleni.

An earlier dome, that of Brunelleschi at Florence (c. 1420), shows the same sort of cracking, although in this case the dome is polygonal, and not a dome of revolution. The cross-section of the dome at each level is octagonal, each of the segments being curved in the 'meridional' direction and flat in the direction at right angles, the 'hoop' direction; the dome is supported by an octagonal drum which in turn covers an octagonal space within the cathedral. A polygonal dome, like a dome of revolution, can be built with a bare minimum of falsework – Brunelleschi was not believed at first when he stated that he could build the dome with virtually no centering. He was secretive and would not reveal his method, saying that if he did, the trustees and expert architects would find it childishly simple. Vasari says that the trustees were won over by the hen's egg, which Brunelleschi passed round the table, challenging those present to make it stand upright. No one could; when it came back to Brunelleschi, he cracked its bottom on the table and so made it stand – a childishly simple solution.

As a matter of interest, a hen's egg has a diameter of say 40 mm and a thickness of 0.4 mm – the 'span-to-thickness ratio' is 100, the same value as that of the fan vaults of Henry VII Chapel, Westminster and of King's College Chapel, Cambridge. By contrast, the same ratio for the Pantheon, and for Brunelleschi's and Michelangelo's domes, is about 10 or a little more, and for a modern large-span reinforced-concrete shell roof, the ratio is about 1000.

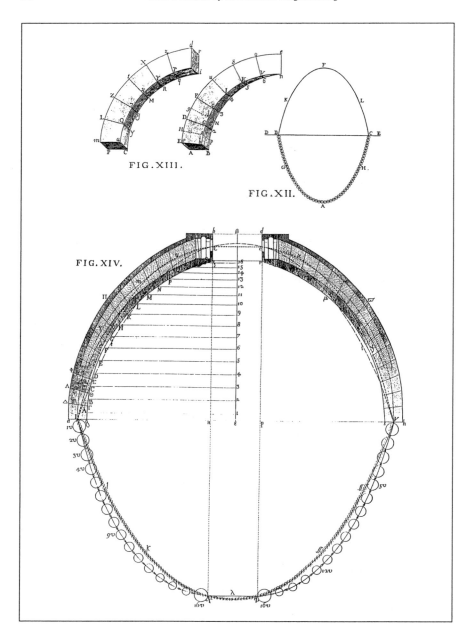

FIG. XIII.

FIG. XII.

FIG. XIV.

Illustrations from Poleni's 1748 report on the cracks
in the dome of St Peter's, Rome.

Giovanni Poleni reported on the cracks apparent in 'Michelangelo's' dome some two hundred years after its completion (by two engineers, Fontana and della Porta, after Michelangelo's death). He starts by reviewing the existing state of knowledge of masonry construction; his scholarship is wide, and he certainly knows of Hooke's hanging chain, FIG. XII. Poleni observes that the existing meridional cracks had divided the dome into portions approximating half spherical lunes (orange slices) – the question to be answered was whether or not these cracks were dangerous.

Poleni states clearly that the necessary condition for stability of the dome is that the forces should lie within the masonry – an explicit foreshadowing of the 'safe theorem'. For the purpose of his analysis he hypothetically slices the dome into 50 lunes, one of which is shown schematically in FIG. XIII (together with half an arch of uniform thickness) – he then proceeds to consider the equilibrium of a complete quasi two-dimensional arch formed by one of these lunes and its reflection. If this arch, tapering to zero thickness at the crown, could be shown to be safe, then the complete dome, cracked or not, would also be safe.

Poleni made the demonstration experimentally. From a drawing of the cross-section of the dome, FIG. XIV, he computed the weight of the sliced arch; for this purpose, each half-arch was divided into 16 sections. He then loaded a flexible string with 32 unequal weights, each weight corresponding to a section of the arch, with due allowance being made for the lantern surmounting the dome. On inversion it is seen that the shape of the chain does indeed lie within the inner and outer surfaces of the dome; the other lines shown in FIG. XIV represent the centre line of the arch and an inverted catenary corresponding to a uniformly loaded chain. Poleni concluded that the observed cracking was not critical, and he agreed with an earlier recommendation that further encircling ties should be provided; the inclination of the hanging chain at its points of support shows once again that there is a horizontal thrust delivered by the sliced arch, and therefore by the complete dome, that must somehow be contained.

The cross-section of St Paul's Cathedral

The triple dome of St Paul's Cathedral consists of a lead-covered timber outer structure, the true conical brick and stone dome supporting the massive lantern, and the inner dome (like the Pantheon, with an eye) which is all that is seen from the inside of the building. Old St Paul's had been so severely damaged in the Great Fire of 1666 that it was pulled down; Christopher Wren's third and final design for the new building was approved in 1675, when building started, although the dome was not constructed until 1705-08, by which time its form had changed significantly from the drawing of 30 years earlier.

Robert Hooke and Christopher Wren were co-surveyors for the rebuilding of London after the Great Fire, and the two worked together closely and amicably for many years until Hooke's death in 1703. Hooke, like Wren, was both a scientist and architect – Hooke's science, but not his architecture, has survived. (For example, Hooke's enormous Bedlam Hospital in Moorfields was begun in 1674, and not pulled down until 1814 when the new Bethlehem Hospital had been built in Southwark. The only surviving works by Hooke in London are two churches, and the Monument, exactly at the north end of Old London Bridge.) Wren knew very well of Hooke's hanging chain; indeed, Hooke's diary entry in June 1675 records that Wren was altering his design of the dome of St Paul's to accord with Hooke's principle. Hooke had in fact in 1671 extended the principle of the two-dimensional arch to the three-dimensional dome of revolution.

The essential difference between Wren's dome for St Paul's and all previous domes lies in the inclined surfaces of the supporting structure – the masonry never becomes vertical, but follows the line of Hooke's (three-dimensional) hanging chain. This inclination is disguised externally by the vertical colonnades of the drum, and is not detectable internally when standing on the floor of the church – but can be seen from the Whispering Gallery.

A dome which has remained virtually uncracked is that of St Paul's Cathedral, London. Wren gave the dome a sophisticated shape in accordance with Hooke's ideas of the 'hanging chain', and he provided encircling chains to prevent any tendency to spread.

Masonry vaults

The semicircular arch requires a massive thickness for stability; the inverted chain can only be contained if the thickness exceeds about 10 per cent of the radius. A semicircular barrel vault to cover the 14 m nave of Amiens would need a thickness, with no factor of safety, of over 700 mm. In practice, circular vaults need not embrace a full semicircle, and thicknesses may then be reduced markedly – moreover, the springings of the vaults, the haunches, may be filled with masonry rubble, adding further to the stability of the construction.

However, the major step in reducing structural weight comes from the use of barrels intersecting at right angles, to create a true three-dimensional vault rather than a repetition of parallel two-dimensional arches. The Romans had used the idea of intersecting barrels to form the groin vault; each concrete bay of the Basilica of Maxentius (AD 30) spans over 25 m. Such a vault needs support only at the corners, and windows may be introduced in the side walls; moreover, smaller thicknesses of vaulting material can be used, so that weights on supporting piers and thrusts on external buttresses are reduced.

Two equal cylindrical barrels will intersect to give a vault which is square on plan, and the diagonals of the square give the location of the groins. If the vault is made of stone (rather than Roman concrete), then geometrical difficulties will be encountered in cutting the stones meeting at the groins – the art of stereotomy is concerned with this problem. There are further severe difficulties if the two barrels intersecting to form the vault have different spans, so that the bay is rectangular rather than square – semicircular barrels of different spans will, of course, have different heights. A first simplification was introduced by allowing the webs of the vault (the severies between the groins) to be slightly domed; the groins were then not fixed uniquely by the intersection of two barrels, but could be designed to some extent by the architect. Once these groins were fixed, the webs could be fashioned with some freedom to fit the boundaries, that is, the groins and the four edges of the bay being vaulted.

A square bay of vaulting

(from T G Jackson, *Gothic architecture...*, 1915).

The vault is formed by the intersection at right angles of two equal pointed cylinders. The diagonals of the square, the groins, indicate the intersections of the cylinders. In late Romanesque, and in Gothic, the groins were constructed first as stone arches, supported by timber centering. The vaults proper, that is, the webs or severies, were then completed by the masons in either the French way or the English way. Both methods lead to robust structures, even if the webs are constructed roughly, with poorly cut stones and much mortar filling the ill-fitting joints. In many cases the finished vault was plastered and decorated, so that its construction could not be seen, but in either case the essential structural action is the same – the groins support the webs, which act almost as two-dimensional arches.

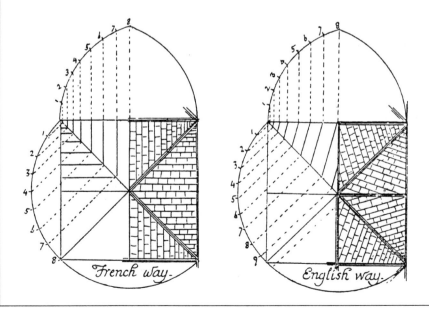

The 'shell' of the Gothic quadripartite rib vault.

The illustration shows a standard Gothic vault as it might be viewed by a modern structural engineer — the actual masonry is idealised into a thin folded structure which can be analysed by twentieth-century shell theory. The mathematics of this theory is on the one hand elegant, and on the other gives a good description of the way the forces are carried — in fact, the mathematics can be discarded as it soon becomes clear that the forces act mainly in the curved directions of the surfaces, and that forces in the 'flat' directions are very small. The vault could, in fact, be sliced into series of parallel arches, just as Poleni had theoretically cut the dome of St Peter's into tapering slices. Viewed in this way, these parallel arches, themselves lightly stressed, depend for support on the diagonal ribs, and these diagonals emerge as the main load-carrying structural elements.

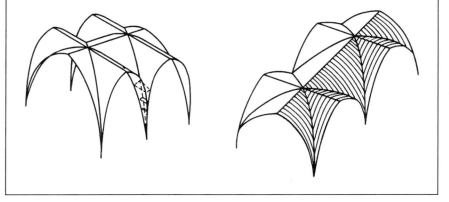

The groins themselves, however, were still difficult to cut, and Romanesque builders finally started their construction of groin vaults by first erecting masonry arches on the diagonals of the bay. These arches were then embedded, either wholly or partially, within the masonry of the vault webs – thus the groins were cut independently, and the rubble masonry webs erected to match.

It was, of course, only a short step to the Gothic vault, with the groin arches built as independent ribs, and the vault webs constructed on the backs of these. The meeting edges of the vault webs could then be cut without the need for a precise match, since irregular joins could be filled with mortar and hidden from view by the groin ribs. The problem of covering a rectangular rather than a square bay was solved by using a pointed vault – the Gothic pointed arch gave rise to the quadripartite rib vault.

The Gothic vault, with its prominent ribs, is easily 'seen' architecturally, whereas a plain groin vault is not clearly articulated. Certainly one of the functions of the ribs is to give a visual definition to the vault, even if the ribs were first introduced for ease of construction. They also have a structural function – two structural surfaces, meeting in a 'crease', require reinforcement at that junction, and the ribs experience a sharp increase in stress compared with that in the neighbouring webs of the vault. Stresses in ribs are still low, however, compared with the crushing strength of the material – if the ribs are not present, then the unreinforced groins will experience higher but still tolerable stresses.

Discussion of the 'function' of ribs is, therefore, complex. They certainly 'collect' forces if they are applied at the intersection of two surfaces. If, however, they are applied to smoothly-turning surfaces (as tierceron and lierne ribs in English Gothic, for example, or on the conoids of fan vaults) then they are decorative – they may have great aesthetic value, but they are not necessary structurally.

There is no evidence that ancient and medieval builders thought in this way about ribs, or, more generally, about the forces in masonry structures. They would, naturally, have had their own ideas which were not concerned solely with geometrical shape and artistic form. An architect who had been through a long apprenticeship, progressing to journeyman and the career grade of master, before being put back to school in the design office to fit him for control of a major work, would himself have experienced the weight of a Gothic-sized block of stone. But he made no calculations.

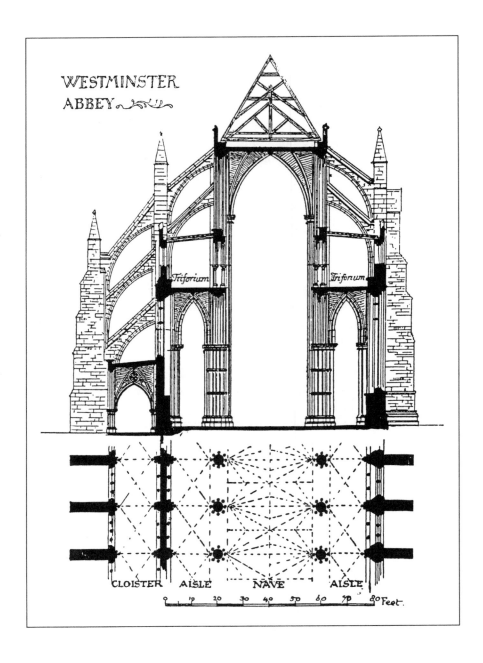

WESTMINSTER ABBEY

Triforium   *Triforium*

CLOISTER   AISLE   NAVE   AISLE

0   10   20   30   40   50   60   70   80 Feet.

The Gothic cathedral – Westminster Abbey

The illustration (from T G Jackson 1906) shows the cross-section of a typical Gothic cathedral. A stone vault covers the high central space, the nave; the nave is flanked by lower side aisles, themselves vaulted in stone. A tall timber roof covers the nave vault – a stone vault lets in water and needs weatherproofing. However, the timber roof is a fire hazard, and the stone provides some protection.

The stone vault thrusts horizontally – if there were no side aisles, the massive external buttresses could be placed directly against the north and south walls of the church to discharge the thrusts to the ground. As it is, the vault thrust is taken over the tops of the side aisles by means of flying buttresses. On the north side of Westminster Abbey the lower of the two flying buttresses props the vault; the upper buttress collects any wind force that may act on the timber roof, and also helps to control the tendency of a timber roof to spread.

On the south side of the Abbey a cloister abuts the wall of the aisle, and the main buttressing pier must be placed yet further away. A relatively slender intermediate pier is introduced, and the two upper flying buttresses form two-span inclined arches. The aisle vault also needs propping; this is accomplished by the main buttressing pier on the north side and by a third level of flying buttresses over the cloister on the south.

The 'mathematical' approach to design came with the Renaissance; the medieval designer's *scientia* was concerned with shapes and proportions rather than stresses and strains.

Chapter 4

Stresses and Strains

The designers of Greek temples or Gothic cathedrals were, rightly, not concerned with stresses – stresses were so small in their buildings that there was, in general, no question of failure of the material. Nor, as a matter of interest, were they interested in deflexions, which was a subject thrown into prominence later with the engineering use of timber, and then the new materials iron and steel. A tree may sway in the wind, but an Egyptian obelisk – Cleopatra's Needle – does not, at least not visibly; nor does the church of Hagia Sofia.

A baulk of timber, however, will not only deflect (perhaps acceptably) under transverse load – if the load is increased enough, the timber will fracture. In modern, and inexact, shorthand, fracture is governed by the attainment of some limiting stress, and the exploration of the unknown territory of stress began in the seventeenth century. The engineering definition of stress is simple and precise, but the notion of stress caused (and continues to cause) some difficulty both in understanding and in application to structural analysis.

There are mathematical difficulties in the description and handling of stress, but a first and fundamental physical problem is that stress cannot be measured. The breaking strength in tension of a stone or timber specimen can certainly be determined, and Galileo in his *Discorsi* of 1638, of which more will be said, shows what appears to be a stone cylinder under test. The weight necessary to cause fracture would give a measure of the 'absolute strength' of the specimen, and Galileo was of course aware that if the specimen were of half the area, then the measured absolute strength would also be halved. Such ideas, which were not, in fact, pursued by

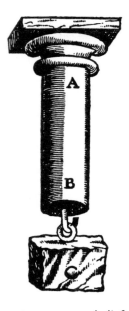

Galileo's *Discorsi*, his *Dialogues Concerning Two New Sciences*, were published in Leyden in 1638. The second new science is concerned with the mechanics of motion; the first gives the first mathematical account of a problem in structural engineering. Galileo wishes to compute the breaking strength of a beam, knowing the strength of the material itself as measured in the tension test shown in the illustration. The drawing does not encourage belief that Galileo ever made such a test (although Galileo himself never saw the illustration – he was blind by the time the book was printed). The hook at B would have pulled out of the stone long before the column as a whole fractured. In the same way, it is thought that Galileo did not in fact drop balls of different weights from the Leaning Tower of Pisa. It is not known that Galileo ever designed crucial experiments of this sort, in order to prove or disprove a theory. What he did was to make crucial *observations*, from which ensued brilliant advances in every subject he touched.

Galileo, lead immediately to the idea of stress, defined as a load per unit area of material.

However, such a definition is purely mathematical – the load can be measured, but the numerical value of stress results only after a numerical operation, namely division by the value of the cross-sectional area of the specimen. What can be measured is deformation – that is, given sufficiently sensitive apparatus, the extension of the specimen can be determined for a given value of the applied load. *Strain* is defined as the extension per unit length of specimen, and is thus non-dimensional – that is, a strain of 1/1000 implies that a bar 1 metre long has extended by 1 mm, and equally that a bar 10 metres long has extended by 10 mm. (A mild steel bar would be near its yield point at a strain of 1/1000.) For a recoverable elastic deformation, Hooke's Law (the same Hooke of the hanging chain) states that stress is proportional to strain – for a given material this constant of proportionality, Young's modulus, may be determined experimentally. Thus if a value of strain is measured, the corresponding value of stress may be deduced simply by multiplying by the value of Young's modulus. It is in fact excessive strain which causes breakdown of a material – steel bars, of different lengths and different cross-sectional sizes and shapes, will all experience distress when the strain exceeds a certain limit (about 1/1000, as noted).

Sophisticated notions such as these were not really explored until the eighteenth and nineteenth centuries. Galileo's enquiry into the breaking behaviour of materials was made against a background of medieval thought, directed to the design of masonry bridges and Gothic cathedrals for which, as has been seen, stresses and strains were very low. For the design of such buildings, ancient and medieval codes gave rules of proportion – geometrical rules which had been found to be effective. If a building were satisfactory, then it would be satisfactory if constructed at twice, or ten times, the size.

Galileo's *Dialogues*

Right at the start of his *Dialogues*, Galileo strikes at the heart of this medieval theory of structural design. Salviati speaks: 'Therefore, Sagredo, give up this opinion which you have held, perhaps along with many other people who have studied mechanics, that machines and structures composed of the same materials and having exactly the same proportions among their parts must be equally (or rather proportionally) disposed to resist (or yield to) external forces and blows. For it can be demonstrated mathematically that the larger ones are always proportionally less resistant than the smaller.'

Galileo's *Dialogues Concerning Two New Sciences* were published in Leyden in 1638, when Galileo was 74. Five years earlier, he had been convicted of heresy, sentenced to life imprisonment and forbidden to publish any more books on any subject. Holland was, of course, outside the reach of the Inquisition, and the Elseviers agreed to publish Galileo's manuscript.

The work is in the form of four dialogues (a fifth was added to the posthumous second edition of 1644). The three Interlocutors are Salviati, who speaks for Galileo; Sagredo, who represents Galileo as a younger man, and who sometimes puts forward views that the older Galileo has rejected; and Simplicio, who might represent a very young Galileo under instruction from the two more learned scientists. Each dialogue is supposed to span a day. The third and fourth parts deal with the development of a science of (pre-Newtonian) mechanics; it is the second day's dialogue that is concerned mainly with structural matters.

However, Sagredo's 'brain is already reeling' at the start of the first day when the mature Galileo, Salviati, mounts his attack on medieval design. He has much more to say about the 'square/cube law' on the second day, but he starts immediately by introducing an example that is, effectively, the archetypal structural problem with which Galileo is concerned. Salviati imagines a horizontal baulk of timber with one end fitted into a vertical wall – the baulk thus acts as a cantilever beam. If the length of the beam is increased there will come a point where it breaks under its own weight, and it is the breaking of cantilever beams that forms the main subject of Galileo's second day of the *Dialogues*.

Immediately after the introduction, however, on the first day, there is a seeming digression. Salviati tells a story concerning a very large marble column that was stored horizontally, being propped on two pieces of timber near its ends. It occured to a workman that the column might break at its middle under its own weight, so he inserted a third prop at the centre. After a few months, the column was found to be broken precisely over the third inserted support. Salviati explains how this had come about. It was found that one of the props near the end of the column had rotted and settled, while the added central prop had remained sound; effectively, then, one half of the column was unsupported. Had the column remained supported by the original two props, all would have been well – if one prop settled, then the column would merely have followed.

This is the famous illustration for Galileo's basic problem – the breaking strength of a beam. Again, the drawing is not really representational, although there is a wealth of circumstantial detail. In this case the hook C may well have been able to carry the load, but the masonry at AB looks insufficient to resist the turning moment at the wall.

This glimpse of the behaviour of what is now known as a hyperstatic structure is not further discussed by Galileo – two props for the column are sufficient, and the third is, in both a technical and usual sense, redundant. The analysis of redundant or hyperstatic structures, is, in fact, the subject matter proper of the theory of structures.

To analyse the fracture of a beam Galileo needs to know the basic strength of the material, and it is for this reason that he determines the 'absolute strength' as given by a tension test. There is no clearer example of the difference in outlook between the scientist and the engineer than the discussion recorded on the first day of the *Dialogues*. An engineer wishes to know only a single number expressing the strength of the material in appropriate units (tons per square inch, Newtons per square millimetre); that number can then be used to determine the fracture strength of the beam. The scientist, on the other hand, wishes to explore the physics of fracture in the tension test. Thus to clarify the discussion, as Galileo says, he imagines a cylindrical specimen of wood, stone, or any other material to be hung vertically, and loaded by an increasing weight at the bottom until it breaks, 'just like a rope'.

Thus starts the series of discussions which make up the substance of the whole of the first day's dialogue; Simplicio can imagine that the longitudinal fibres extending the whole length of a wooden specimen can make the whole specimen strong, but asks how a rope, composed of fibres only two or three braccia long, can be equally strong. Salviati explains how the short fibres are twisted together to form a long rope, their mutual interaction conferring strength on the whole.

But this does not help to explain the fracture of apparently amorphous materials, such as marble, metal or glass, and Salviati admits the difficulty of this problem. He has the view that the particles of a body have some inherent tenacity, and he also refers to the well-known repugnance that nature exhibits towards a vacuum. The question of voids arises from the experience that two slabs of material, if smoothed, polished and cleaned, may be slid one against another but are difficult to pull apart. Thus an apparent solid might consist of a very large number of small parts having inherent strength, and also of a very large number of voids adding to the strength. Indeed, the number of voids could be infinite, and an hour or so of discussion is devoted to the mathematics of this idea – the proof that a finite area, for example, can contain an infinite number of voids.

One mathematical topic leads naturally to another, and Salviati, the mature Galileo, is constantly reminded of interesting results, to the point where Sagredo is forced to remind him how far they have strayed from their subject. But Salviati has been concerned with trying to explain how expansion and contraction (due to temperature changes, for example) can occur without assuming interpenetration of bodies or the introduction of voids and the discussion turns, almost inevitably, to Aristotelian theories of motion. Motion is the second new science, exposed on the third day, but it is on the first day that Galileo makes the famous statement that two bodies of different weights will fall under gravity at the same speed. Simplicio finds this hard to believe, although Sagredo says he has made the test. Whether he did or not, it is clear that a real test is being reported; Salviati, explaining to Simplicio, notes that Aristotle says: 'A hundred-pound ball falling from the height of a hundred braccia hits the ground before one of just one pound has descended a single braccio.' In fact, they arrive at the same time, or rather (and this is the evidence of a test) the larger is ahead by two inches because of the effect of air resistance. There follows a long discussion of motion in dense and rarified media, water and air; and then of the oscillation of the pendulum; and so to the vibration of musical strings. Even Salviati, as night falls at the end of the first day, wonders how the discussion has been carried on for so many hours without tackling the main problem.

The problem is stated again clearly at the dawn of the second day – it is to find the strength of a beam when it is broken as a cantilever. It is evident that fracture will occur at the root of the cantilever, where it is embedded in the wall. At this embedment, the beam will develop its 'absolute strength', resisting the turning moment of the applied load, and Galileo quickly shows that the bending strength of the beam is equal to its absolute strength in tension multiplied by half its depth. Galileo has reached the correct result that the strength of, say, a rectangular beam is proportional to the width of the beam and to the square of its depth.

It turns out, however, that Galileo's value of $\frac{1}{2}$ for the constant of proportionality is not necessarily correct – the idea that all the 'fibres' at the root of the cantilever can be mobilized to give their full breaking strengths is, in general, wrong. The absolute strength cannot be introduced in this way. However, Galileo made completely correct deductions from his analysis – for example, that the bending strength of a circular cylindrical beam is proportional to the cube of the diameter. Other examples follow, some of

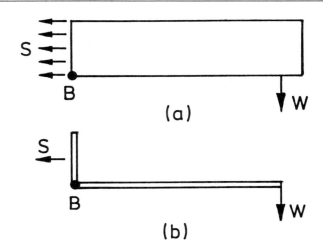

Galileo's explanation of the breaking of a cantilever beam.

The beam is fixed in the masonry wall, and bears on the masonry at point B. As the load W is increased, a stage will be reached at which the 'absolute strength' of the beam will be mobilized — this strength is given by the breaking load of the beam tested vertically as a column. Thus in Fig. (a) the absolute strength S resists the moment about B of the load W; Fig. (b) shows the equivalent 'bell-crank' from which, knowing the dimensions of the beam and the value S of the absolute strength, the value W of the breaking load may be calculated. The calculations were not simple in 1638; Galileo uses Aristotle's laws of the lever (better stated by Archimedes, says Galileo) to solve the problem of the bent bell-crank. Diagrams such as Figs (a) and (b) were not drawn by Galileo, but his calculations were, of course, correct, although his final result contained a debatable numerical factor.

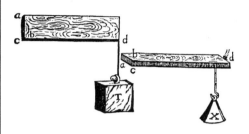

Galileo's illustration of a result from his calculation of the strength of a rectangular beam. Although he obtained a wrong value for a numerical constant, this is not of consequence if *ratios* of strength are discussed. If a plank is set flat and on edge, then the ratio of strengths in the two cases is simply the ratio of the lengths of the sides of the rectangle — a 200 × 25 mm plank is 8 times as strong when set vertically rather than horizontally.

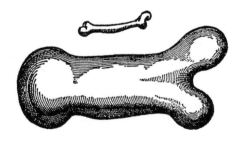

Galileo's sketch of two bones corresponding to animals one of which is three times the linear size of the other. Their weights are thus in the ratio of 27 to 1, and the gross size of the larger bone is necessary if the two bones are to perform the same function. Elephants are mentioned as examples of the largest animals that can exist on land, and Simplicio notes that whales ('ten times the size of elephants') are in effect weightless when immersed in water.

which demonstrate the square/cube law for structures subjected to gravity loading; the loads are proportional to the cube of the size, whereas the supporting areas are proportional only to the square. This culminates in Salviati's general statement about the impossibility of building enormous ships, palaces and temples – 'nor could nature make trees of immeasurable size, because their branches would eventually fail of their own weight'. The point is carried through with a study of comparative anatomy.

The bending problem

Galileo's problem of the breaking load of a cantilever beam appears to be concerned with the behaviour of a complete structure, but it is in fact one of the simpler examples of a calculation in the field of *strength of materials*. This branch of the structural engineer's science is concerned with the close examination of a small region of the structure – in this case, the embedded root of the cantilever – to check whether or not the conditions in that region are satisfactory. That is, given the results of a simple tension test, which will provide (in modern terms) a value of the yield (or fracture) stress of the material, how is this related to yield (or fracture) at the critical section? Galileo noted that the tip load on the cantilever produced bending in the beam, and this bending was resisted, at the root of the cantilever, by the mobilization of the 'absolute strength' of the cross-section. A simple equation then gives the value of the breaking load, and, by implication (and, indeed, stated explicitly by Galileo), a smaller load will not break the beam. A modern engineer, designing such a beam for a given load, will apply a *factor of safety*, which will have different values under different circumstances, but typically might be about 2 – a beam to carry a working load of 5 tonnes might be designed to collapse at 10 tonnes.

   With two major exceptions, examination of local conditions in a structure (that is, problems in strength of materials) formed the area of study of structural scientists for the next two hundred years. One of the exceptions was the treatment of the masonry structure, and in particular the arch, which was discussed in Chapter 3; the other sprang from the realization that members of a structure could bend, and might, under certain conditions, become unstable. Buckling is considered in Chapter 5. Galileo had, however, started a line of enquiry that was to be followed by very many other workers.

There is no record that Galileo made any bending tests to confirm his calculation of the breaking load of a cantilever beam. Others did, however, make such tests soon after Galileo's theory had been published, and there was considerable discussion of the problem before the end of the seventeenth century, and indeed through the next into the nineteenth. Mariotte in France, for example, in his book of 1686, records tests in both tension and bending – he could not relate the results of the two by Galileo's formula. He concluded that Galileo's theory must be wrong, and he re-examined the analysis.

Like Galileo, Mariotte assumed that when the cantilever beam fractured it would turn about its lowest point where it was embedded in the wall; however, he imagined that, as one moved up the wall, the fibres would be loaded more and more heavily. Instead of a uniform distribution of stress (the mobilization of the absolute strength of the section), Mariotte took a linear distribution. Mariotte deduced that a beam was less strong than Galileo had predicted, and his own formula gave reasonable agreement with his own experiments.

Agreement between theory and experiment was in fact only approximate; some giant interventions – those of Leibniz and James Bernoulli, for example – covered no new ground, but testified at least to the importance of the problem. Parent, however, in 1713, spotted a flaw in the theories under discussion – both Galileo's uniform distribution of stress, and Mariotte's triangular distribution, summed to a horizontal pull acting on the end of the cantilever beam. No such pull was supposed to be present in the theory, nor had a pull been applied in the tests. Parent concluded that one should imagine that the top fibres of the beam were extended while the bottom fibres were compressed – it would then be possible to arrange the forces so that there was no net push or pull at the root of the cantilever. If these forces varied linearly over the depth of the section (and Parent saw that there was in fact no need to assume a linear distribution) then the new calculation of bending strength gave a coefficient of $\frac{1}{6}$ instead of the $\frac{1}{2}$ of Galileo or $\frac{1}{3}$ of Mariotte. For the first time, a logical and correct mathematical description had been given of the way a beam might fracture, but in fact none of the three values of coefficient accorded with tests – $\frac{1}{2}$ might be better for stone and $\frac{1}{3}$ for wood, while the mathematically correct value of $\frac{1}{6}$ seemed to be useless as a predictor of fracture.

It was shortly after Parent's work that the first comprehensive 'standard code of practice for civil engineering' was published – Bélidor's *Science des*

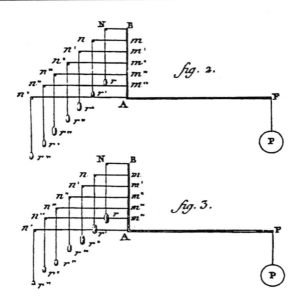

The figures are from a book by Girard published in Year 6 of the Revolution (1798). They give visual interpretations of the conditions at fracture of a cantilever beam – Fig. 2 is Galileo's theory, in which all fibres of the cross-section are stressed equally (represented by equal weights acting over pulleys), and Fig. 3 shows Mariotte's theory, where the fibres contribute stresses in proportion to their distance from the fulcrum A. Although Galileo had introduced the idea of imaginary 'fibres' in a solid (such as stone or wood), all his fibres were stressed equally – the fact that the 'stresses' might vary through the depth of a beam is a highly complex idea. Calculations based upon the two theories agree that the strength of a rectangular section beam is proportional to the width and the square of the depth, but Galileo has a factor of $\frac{1}{2}$ and Mariotte $\frac{1}{3}$. Girard in 1798 was of the opinion that Galileo's theory was best for stone, and Mariotte's for wood.

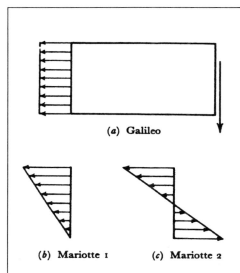

(a) Galileo

(b) Mariotte 1   **(c) Mariotte 2**

The figure shows alternative assumptions of the way stresses might act on the end of a cantilever beam at the point of failure. Figure (a) corresponds to Galileo's use (1638) of the idea of 'absolute strength', and Fig. (b) to Mariotte's linear distribution (1686). Both of these distributions sum to a horizontal pull on the end of the beam, which is not present in theory or practice. Mariotte had in fact considered the distribution of Fig. (c), in which the pull on the top half of the section is exactly balanced by a push on the bottom half – however, he made an arithmetical mistake in his working, and did not pursue this idea. Parent (1713) showed that it was necessary to have some stress distribution such as Fig. (c) if equilibrium were to be satisfied – no horizontal pull on the beam.

ingénieurs appeared in 6 books in 1729, and was influential for the rest of the century. Bélidor deals with problems of geotechnical engineering (soil thrust), with the design of arches, with properties of materials, and with the drawing up of specifications and contracts. Bélidor's handbook is an up-to-date 'Vitruvius', and even the form shows similarities, with one of the six books, for example, given up to a description of the classical Orders, taper and entasis of columns, and so on. It is an early example in modern times of an engineer distilling the findings of science into rules of design, with no real attempt to present any theoretical basis for those rules. Bélidor discusses also the strength of beams, and knows of Parent's early work; he offers no new theory, but presents practical design rules which effectively held sway for 50 years, until Coulomb wrote his first scientific paper.

One of Coulomb's four problems of 1773 was the fracture of beams; he had used Bélidor's *Science des ingénieurs* as a textbook at his university, and he knew of the work of Galileo, but, writing in isolation on the island of Martinique, he reinvented much theory that had already been discovered by others. In fact, Coulomb makes exceptionally clear statements of the problem; nobody had noticed Parent's analysis, which had been published in small thick volumes, discouraging to the reader, so that Coulomb's attack on the bending problem was in the nature of a true rediscovery. Coulomb himself was a great experimenter, and he had found that stone and wood behaved in very different ways. He presents two theories, one for each of the materials, but based on a common approach that satisfied the requirements of mechanics (no pull on the beam, for example). He concludes that Galileo's coefficient of $\frac{1}{2}$ seemed best for stone, but did not verify his (and Parent's) coefficient of $\frac{1}{6}$ for timber against his own experiments. At the turn of the century, just after the Revolution and the Reign of Terror, the standard text of the *École Polytechnique* (founded 1794) stuck to Galileo's formula for stone, and to Mariotte's (coefficient of $\frac{1}{3}$) for wood.

Navier and the *Ponts et Chaussées*

The (Parent)/Coulomb theory of bending became the Coulomb/Navier theory. Claude Louis Marie Henri Navier (1785–1836) attended the *École Polytechnique* from 1802 to 1804, and then transferred to the *École des Ponts et Chaussées*. On graduation, he became an *Ingénieur des Ponts et Chaussées*, and finally in turn taught at the school. In 1813, he edited a new revision of Bélidor's book, and he soon started to publish scientific papers in many structural fields, including the bending of plates, the theory of suspension bridges and so on. However, perhaps his most influential work is to be found in the publication in 1826 of his lecture notes, the *Résumé des Leçons données à l'École des Ponts et Chaussées*. These notes still contain some errors of theory; the great elastician Barré de Saint-Venant edited Navier in 1864, and he corrected and expanded the text enormously by footnotes. Among other matters, Navier did not have clear ideas about the way to treat shear stress – this is discussed below.

The importance of Navier's 1826 *Leçons* lies in the great sweep of subjects covered. Not only are problems in the field of strength of materials discussed – for example, local failure in bending (Galileo's problem) –

buckling theory is also exposed (Chapter 5), and a general theory is developed to handle hyperstatic structures (Chapter 6) which is the start of *theory of structures* proper. The book is effectively the first modern text on structural analysis, in which scientific theory is directed towards the determination of the sizes of structural members in order to perform specified structural tasks.

There is a unifying idea which is common to all of Navier's analysis. Navier states, implicitly, that the engineer is not interested in the final, ultimate state of a structure in which it collapses – the intention of the engineer is to prevent collapse. Thus a Galilean calculation of the breaking strength of a beam is not of the right kind – the engineer is concerned to ensure the safety of a structure under its specified loads. To this end, the engineer must calculate the stresses in the structure under those loads, and ensure that the stresses are below the elastic limit of the material.

Thus a philosophy of design emerges and is crystallized in Navier's writings of 1826. The linear theory of bending, formulated by Mariotte and developed by Parent and Coulomb, is interpreted physically, by reference to Hooke's Law, as a linear elastic theory. Elastic deformations are recoverable; a structure loaded and then unloaded will suffer no permanent 'set'. Moreover all the equations are linear – a doubling of the load (within the elastic limits) will double the deflexions. Stone, in fact, is much stiffer than timber, and was often regarded as 'rigid'; more correctly, stone, compared with timber at least, is brittle, and does indeed behave in a more or less linear elastic way as it is loaded up to its breaking point. By contrast, timber has more 'give'; not only are structural deformations more evident, but there is some 'ductility' – the behaviour becomes non-linear above a not well-defined limit of stress.

In addition to these two materials, iron was coming into use structurally in Navier's time. Cast iron too is brittle, with fracture occurring soon after the elastic limit has been exceeded; wrought iron is more ductile, and can accommodate a small amount of permanent set without breaking. It was now becoming commonplace to test the strengths of these new materials, as well as of the old, and the newly formulated elastic theory could be applied as Navier proposed – calculated stresses at a critical section of a structure should not exceed a certain proportion of the elastic limit. Breaking loads were not entirely forgotten – indeed, tests on the very newest of structural materials, mild steel, which came into use in the late nineteenth century, showed that it was hard to cause fracture even at very large strains. A steel (or wrought-iron) rod may be bent permanently and develop a kink at the critical section – the so-called plastic hinge in bending.

This kind of behaviour had been considered in a theoretical way by Saint-Venant in his 1864 edition of Navier. He postulated a general non-linear distribution of bending stress and, by manipulation of the constants in the equations, showed that all theories could be embraced, from Galileo through Mariotte to Coulomb. All that was necessary was to make some experiments, and then to assign empirical values of the constants in the formulae to accord with the results of those experiments. The engineer would then be in a position to predict the consequences of any forces acting at a section of the structure.

However, it was the linear elastic theory which was the simplest, and which became dominant in structural design. The further development of structural analysis was to take place within the constriction of the strait-jacket imposed by Navier and the great *Écoles*.

Shear stress

The tip load acting on Galileo's cantilever causes bending of the beam throughout its length; the value of the bending moment is greatest at the critical section, the root of the cantilever, and this is where fracture will occur. In the working state, when the section is still elastic (as Navier would wish), then the upper fibres of the beam are in tension, and the bottom fibres in compression, with the bending stresses changing smoothly from one state to the other through the depth of the beam, being zero on the centre line. All this emerged from the correct mathematical approach of Parent, Coulomb and Navier himself. The cross-section of the beam was usually, by implication, taken as rectangular, although Galileo and some later workers applied the theory to circular wooden logs.

Engineers like Brunel, in the mid-nineteenth century, who were introducing iron into their structures, realised that material near the centre of the beam, more lightly stressed than the material at the surfaces, was to some extent 'wasted', and they developed the I-section plate girder. Here masses of metal are concentrated at top and bottom of the girder, and these flange plates are connected (by means of angles and rivets or bolts, and later by welding) to a third plate, the web. Thus the flanges, the top

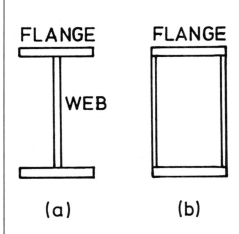

FLANGE

WEB

(a)

FLANGE

(b)

The I-section girder, Fig. (a), was developed to make economical use of material. The two flanges, at the top and bottom surfaces of the beam, are positioned to be most effective in resisting bending. Vertical loads – shear forces – must also be carried by the beam and transmitted from section to section along its length; these 'dragging' forces are carried largely by the web of the I-section. For usual civil engineering use, the two calculations – for the flanges in bending and the web in shear – can be kept separate. The I-section is, however, weak when loaded laterally rather than vertically; even if lateral loads are not present, the section is liable to buckle sideways under purely vertical load. A much more stable configuration, using the same amount of material, results from splitting the web in half to form the box-section, shown in Fig. (b). The simplified basic structural calculations for the two sections are exactly the same, but it turns out that in fact the shear stresses in the box-section 'flow' into the flanges, whereas the I-section flanges are virtually free of shear. This was known to aeronautical engineers, but was not at first appreciated by engineers designing conventional land-based structures – it was the root cause of collapse of the first box-girder bridges in Australia, Germany and the UK.

in tension and the bottom in compression, provide the bending strength of the section, and the web holds the two flanges apart to maintain a stable structural form.

The web in an I-section girder performs, however, another structural function. The tip load on the cantilever, which causes bending, must itself be carried from section to section along the length of the beam, and it finally imposes a downward force on the supporting structure at the root of the cantilever. This vertical force is carried in shear – each imaginary vertical section of the beam drags down on its neighbour. The effect is present in the beam of rectangular cross-section, and was remarked for what seems to be the first time by Coulomb; it is, in fact, of little consequence for such a beam, and was explicitly ignored by Coulomb. The web in an I-section girder performs the same shearing function in transferring the vertical forces from section to section, but its design is now of some significance. The web must not be too thin, for example, or it might fail under the action of high shear stresses induced by the vertical loads.

Shear stress, a dragging stress on the cross-section, seems to be of a different kind from the pushes and pulls on the section induced by bending. Indeed, the analysis of stress is mathematically complex – Navier was unable to make this analysis in 1826, although the mathematicians had begun to consider the problem at that time. It fell to Saint-Venant in his 1864 edition of Navier to sort the matter out as far as engineering design was concerned, and simple rules emerged for the design of I-section girders. These rules were eventually embodied in design manuals, and engineers could quickly size up the steelwork necessary for civil engineering structures.

Different kinds of progress were made in the twentieth century by aeronautical engineers. The structural members for aircraft had to be as slender as possible if the aircraft were to fly at all; thicknesses of sections were reduced to the minimum, and aluminium alloys, more expensive but lighter than steel, and as strong, were exploited. The sections had thin walls, and it was quickly found that a box-section gave economical components. It was important for the design of these members that an accurate analysis of stress should be made, and it turned out that the distribution of shear stress was markedly different from that of the I-section girder. The shear stresses are no longer confined mainly to the web, but can cause high compressive forces in the flanges of the section.

Chapter 5

Flexure and Buckling

Galileo did not speculate on the *shape* that his cantilever might assume when loaded at the tip, although others were examining this question before the end of the seventeenth century. The problem may not have presented itself as important – and indeed, from some viewpoints it is not – but certainly the problem was difficult to solve with the mathematical tools available at the time. Some curves had been known since the time of the Greeks – the conic sections, for example: that is, the ellipse, parabola and hyperbola – but there was no general mathematics available to describe more complex shapes. The celebrated dispute as to whether Newton or Leibniz had invented the calculus came right at the end of the century; it was the lack of knowledge of this mathematics that prevented Hooke in 1675 from finding the shape of the hanging chain. Thus Pardies (for example) in 1673 had a clear understanding of the basic laws of statics – the law of the lever, forces in windlasses, gear trains and so on – but he had to assert, without mathematical proof, that the bent shape of the tip-loaded cantilever was a parabola. This seems to be the first discussion of the problem, and Pardies' assertion is wrong.

Newton's method of fluxions – the calculus – was well-developed by about 1670; with the method, he could discuss problems of motion – the relation between velocity and distance, for example. The method had precedents; Cavalieri in 1635 gave mathematical results which were useful in mechanics. And Galileo's second new science of 1638 was precisely concerned with the study of motion. The calculus, however, was broader than these forerunners, and enabled mathematicians to tackle any problems in which the quantities under discussion were varying continuously – in space as well as in time. For Galileo's cantilever, for example, the 'bending moment' in the beam increases linearly with distance from the tip load, so that each cross-section is subject to a bending action which reaches its maximum at the root. A straight strip of wood in 'pure bending' – held in the hands at each end and bent equally by each hand – must clearly take the form of an arc of a circle. But what is the shape of such a strip, Galileo's beam, subject to variable bending?

James Bernoulli solved the problem in 1691 by the use of the calculus. His physical master statement – sufficiently powerful to be hidden initially in the form of a Latin logogriph – was that the curvature of the arc of the circle in pure bending (that is, the inverse of the radius of the circle) is proportional to the value of the bending moment. For each infinitesimal portion of Galileo's beam, at which the bending action of the load is known, then the local radius of curvature is also known; the overall shape will be given by 'integration' along the beam. It turns out that the equations are easy enough to write, but that their solution presents formidable problems. James Bernoulli was not tackling an engineering problem but a mathematical one, and he wished to obtain an exact expression for the shape of the deflected beam, no matter how large those deflexions were.

James's nephew Daniel, some 50 years later, saw that an 'engineering' approach would simplify the mathematics considerably. A beam, if it is to be of practical use in a building, will suffer very small deflexions – small, that is, compared with the overall dimensions of the structure. Indeed, deflexions are often so small that they are not critical criteria for structural design; a masonry cathedral does not deform visibly under snow or wind loading, and Galileo was right to concentrate, for the sort of structure he was considering, on strength rather than stiffness. As will be seen, however, Navier's formulation of the structural design process requires the ability to calculate deformations, even if those deformations are of negligible importance in the functioning of the structure. However, a century before this, Daniel Bernoulli saw that if deflexions of a beam were indeed small, then awkward terms could be crossed out in the basic equations. These equations then solve easily; for all practical purposes, the shape of Galileo's cantilever under a tip load turns out to be a cubic curve (and not the quadratic curve, i.e. the parabola).

As a matter of interest, Daniel Bernoulli was able to solve the much more difficult problem of the determination of the frequency of vibration of Galileo's cantilever – one arm of a tuning fork, say. Moreover, he determined not only the fundamental frequency, but also the overtones, and he showed that the overtone frequencies are irrational with respect to the fundamental. That is, the overtone frequencies can never be simple multiples of the basic frequency, and the composite note emitted by a tuning fork must always be dissonant. Daniel Bernoulli made experiments to verify his theory of vibration.

Leonhard Euler

Daniel Bernoulli is best known for his work on fluid mechanics; 'Bernoulli's Equation' is fundamental to the study of flow problems. He continued to work in the field of solid mechanics, however, and he noted a curious property of elastic beams. When such a beam is bent from the straight, energy is stored (and can be recovered on removal of the load); under load, the elastic beam takes up a shape such that that strain energy stored is a minimum. This property was later to assume almost mystical qualities, with engineers formulating the idea that Nature responded to environmental attack with 'least work'. Daniel Bernoulli himself, however, formulated strain energy in purely mathematical terms.

The infinitesimal calculus of Leibniz and Newton had developed rapidly in the hands of James Bernoulli and his brother John – John was the father of Daniel. The Bernoullis, originally from Antwerp, had been settled for several generations in Basel, and Leonhard Euler, a native Swiss, was one of John Bernoulli's pupils. Euler was soon to be recognised as the outstanding mathematician of the eighteenth century; his textbooks on differential and integral calculus influenced all the mathematicians of the time.

Euler had developed a new branch of calculus, the calculus of variations, and Daniel Bernoulli wrote to him with the challenge of solving the problem of the 'elastica' with his new invention: A uniform elastic strip of fixed length is required to start from a given point (with a given slope at that point) and to end similarly at another point, also with a known position and direction; what should be the shape of the strip to make the elastic strain energy as small as possible?

Euler rose to the challenge in 1744 by adding an appendix to a new book on the calculus of variations. He easily obtained the relevant highly complex equation and proceeded, purely mathematically, to transform it to a state from which he could make physical deductions. With a bare

The calculus of variations deals with problems such as the following. (1) A piece of string of given length is to be laid out to enclose the largest possible area; what is the shape of this area? This was Dido's problem, when she was promised as much land to found the citadel of Byrsa in Carthage as could lie within a bull's hide – she cut the hide into strips, tied together to form one long string. The answer, a circle, is perhaps obvious, but the formal proof requires the calculus of variations. (2) A bead slides freely under gravity along a smooth wire from a given point A to a lower given point B (where B is not directly below A). What should be the shape of the wire, if the journey is to be as short as possible? The 'obvious' answer, a straight line, is in this case wrong; James and John Bernoulli were the first to show that the curve should be a cycloid.

In problems like these, some quantity (area enclosed, time of descent) has to be maximised (or minimised) subject to certain restraints (fixed length of string, given points A and B). The calculus of variations generates a mathematical equation expressing the solution (the circle, the cycloid) for such a problem.

minimum of calculation, at this stage, Euler was able to sketch all possible shapes that could be taken by the bent strip; finally, he made heavy calculations to give numerical results. There are no approximations in this work, apart from those involved in rounding off the numerical calculations, and the elastica can assume fantastic shapes, seemingly far removed from any possible engineering application.

Euler himself, however, paid particular attention to his Class 1, in which the initially straight elastic strip is bent only infinitesimally. He found that this very small displacment was that of a (half) sine wave and, as a surprising observation of outstanding importance, the displaced shape could only be maintained in the presence of a particular load of calculable value. If the elastic strip is taken to represent a column in a building frame, then the particular load represents a vertical compression of the column.

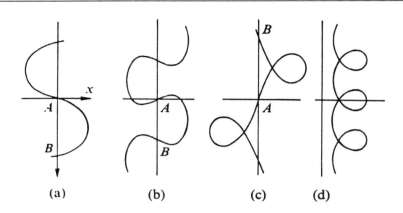

(a)   (b)   (c)   (d)

Possible shapes for the elastica – the curve taken by a strip of fixed length when bent from the straight, and required to connect two given points with specified slopes at those points. Euler distinguished nine different classes, of which the second, fourth, sixth and eighth are shown. The other classes are special cases; Class 5, for example, represents the transition between Figs (b) and (c), and is a figure 8 lying on its side. Class 1, of particular importance to engineers, is of the form of Fig. (a) but with only infinitesimal deformation from the straight.

This statement can be written the other way round, and Euler himself was at once aware of the application – a vertical column can sustain comfortably an axial load of any value up to a certain limit, where that limit is precisely calculable. At this limiting value, which is now known as the Euler buckling load, the column will deflect sideways and become unserviceable as a structural element. Euler's calculation predicts that the buckling load is inversely proportional to the square of the length of the column – two columns of the same cross-section, and one half the length of the other, will have buckling loads in the ratio 4 to 1. This inverse square law had actually been found experimentally by Musschenbroek a few years earlier.

Buckling behaviour is not a problem with traditional masonry construction; the column of a Greek temple, for example, is working at a load far below its 'Euler' buckling value. Timber sections, used vertically as columns rather than horizontally as beams, are more critical, since their dimensions are more slender than those of corresponding masonry elements. As has been noted, iron was being introduced into structural work, and it was the exploration of the properties of these 'new' structural materials that concerned Musschenbroek, and others, in the eighteenth and nineteenth centuries. Their experimental results, and Euler's theory, were immediately taken up in the teaching and practice of structural design – Navier's 1826 text makes full use of the work.

Euler's first study was based upon a general discussion of the elastica, with large deflexions being considered. Euler himself, however, having realised the importance of the practical application of ideas of buckling, reworked his analysis in the way Daniel Bernoulli had done for Galileo's cantilever. That is, instead of writing general equations, and then specialising the solutions to apply to the engineering design of columns, Euler assumed from the start of his new analysis that deflexions would be small, so that he could make a corresponding great simplification of the equations.

Buckling

With the introduction of hard-working strong materials fabricated into economical slender elements, buckling has become of increasing importance in design. The lateral buckling of columns represents almost the simplest case; even so, calculation becomes more difficult when attention is paid to the way the column is connected to the rest of the structure. Are the ends of the column restrained by the beams it supports – or, on the contrary, do those beams impose bending on the column; is the top of the column free to move sideways; are the bases of the columns connected to rigid foundations? These are some of the problems which must be resolved in the formulation of practical design rules.

However, there are other forms of structural buckling, some of which lead to heavy problems of analysis, but all betraying the same kind of behaviour. A load is applied to a structure, or to one of its components, and is slowly increased in intensity. Although internal forces in the structure will increase, and displacements may perhaps be measured, the structure as a whole may well appear unchanged; at a certain level of loading, however, some unexpected development occurs, analagous to the 'Euler' buckling of a column. Galileo's cantilever beam, for example, if it were made from a steel I-section, or indeed from a thin deep plank of wood, may show no untoward behaviour as the tip load is steadily increased; at a certain value of tip load, however, and before any limiting stress has been reached at the root of the cantilever, the beam may well move out of its own plane, and not only deflect sideways, but twist. Such behaviour is liable to induce large strains at the root, not accounted for by the simple bending theory of Galileo/Coulomb/Navier; although the initial buckling movements may be quasi-static, the large strains will lead to failure of the material and catastrophic collapse of the structure.

In practice, the onset of buckling will not be precisely observable, although the theoretical load at which buckling occurs may be calculated accurately. A practical structural member – a column in a building frame, for example – is never exactly straight, nor is a column load ever truly axial. As a result, a real column subjected to increasing load will show small increasing deflexions from the start, but these deflexions do not become significantly large until the theoretical buckling load – the 'Euler load' – is approached. Indeed, the theoretical limit cannot be quite reached before catastrophic collapse occurs.

The word 'catastrophe' obviously implies a situation that should be avoided in structural design. Some protection will be afforded if the structural material has some ductility, and is capable of sustaining a measure of overstrain without rupture. Glass, or cast iron, are poor materials in this respect, and good scientific reasons (rather than instinct) for their avoidance in practice are noted in Chapter 7. Mild steel, however, as used in rolled steel joists or larger fabricated sections, and to some extent wrought iron, or aluminium alloys, or timber, or (as it turns out) reinforced concrete, all possess a good capacity to tolerate overstrain, and a mere 'kinking' of a member will not imply that the whole structure is at risk.

Buckling behaviour, however, is 'brittle', even if the material involved is itself ductile. As has been noted, very little appears to happen to a mild steel column until the load approaches the 'Euler' buckling value; at that level, however, large lateral deflexions will quickly lead to yielding of the material, and collapse of the whole structure may be promoted. In whatever way a designer may choose to calculate the forces on a structure, and whatever method is used to assign sizes to its components, buckling must be avoided.

Euler's work on the simple column has been extended over the last 200 years to cover many cases of practical importance, for example lateral-torsional buckling of beams in building frames, or the wrinkling of thin skins in aircraft structures (which may well be a stable and non-catastrophic form of buckling). The theory is difficult, and approximations have to be made; moreover, the calculations are often very sensitive to practical imperfections of manufacture and assembly. A saving grace is that, even if estimates of buckling loads cannot be given firmly, these estimates are improved markedly by quite small changes in dimensions of the members. An increase in column size in a building frame, for example, or a slight thickening of the material in an aircraft boom, will give calculated loads which the designer can consider to be completely safe when compared with the likely practical conditions.

Chapter 6

The Theory of Structures

It was clear to Galileo that a beam resting on three supports could be subjected to forces not envisaged by the engineer. Two supports are sufficient to carry the beam and, for a known weight, the use of simple equations of statics (well understood by Galileo) enables the support forces to be found. This simple structure is, in fact, *statically determinate*; once the support forces are known, further straightforward equations enable the internal forces in the beam to be calculated, and the science of strength of materials, summarised in Chapter 4, may be used to assess the suitability of the dimensions of the beam.

By contrast, the introduction of a third support for the beam makes the structure *statically indeterminate*, or *hyperstatic*; there is no way that the simple equations of statics can be used on their own to find the propping forces at the three supports. Technically there are available only two relevant equations, and these do not suffice to determine three unknown quantities. Galileo did not himself explore this idea, nor did he give any general analysis of the hyperstatic structure. Instead, in his account of the broken column he pursued, very briefly, a different line of discussion, in which he considered displacements, rather than forces. The line proved, much later, to give an alternative and powerful method of attack on the problem of the analysis of structures. Galileo's explanation of the breakage of the column stored horizontally sprang from his understanding that no such accident could happen to a beam resting on two supports only. Should one of the supports of this statically determinate beam settle, by decay or otherwise, then indeed it would move slightly, but the forces to which it was subject would not alter – the statics of a statically determinate beam are unique. By contrast, decay of any one of the three supports of the hyperstatic beam, with a consequent small, almost infinitesimal, settlement of that support, will alter radically the force system acting on the beam, and hence alter also the internal forces in the beam. In Galileo's account, the bending moment at the inserted central support became so great that the beam fractured there.

It is the calculation of internal and external forces that arise in response to given loading on a hyperstatic structure that forms the subject matter of the theory of structures.

Work before Navier

Thus Galileo, while having in a sense founded the science of the theory of structures, made no contribution to the subject proper. His concern was with the breaking strength of a statically determinate cantilever beam, and he explored completely new ground in applying mathematics to the solution of this problem. As has been seen, experiments by others (for example, Mariotte in the seventeenth century, and others throughout the eighteenth century) showed that the *form* of Galileo's results was correct, but that numerical adjustments had to be made if the theory were to be correlated with experimental results.

Mariotte tested 'simply supported' beams, that is, statically determinate beams on two end supports. He also tested some fixed-ended beams – the ends of identical specimens were inserted in mortises so that they were fixed both in position and against rotation. Such a 'clamped' beam is hyperstatic – the clamping forces at the ends cannot be found from the statical equations alone. However, Mariotte was not troubled by any such notions – he noted that the beam collapsed when fracture occurred at the centre (as for the simply supported beam) and also, as might be expected, at the ends, where the beam was located in the more or less rigid mortises. In this collapse state, the beam does in fact become statically determinate, which is one of the key consequences of modern plastic theory, to be discussed in Chapter 7. Mariotte found that the collapse load of a fixed-ended beam was just double that of the corresponding simply supported beam, a result exactly in accordance with the predictions of plastic theory.

A similar but theoretical collapse analysis was made by Girard right at the end of the eighteenth century, and he exposed clearly the way in which the equations of statics might be deployed in order to solve the problem. Girard also explored what was, in effect, Galileo's hyperstatic problem – the beam on three supports – indeed, Girard extended his analysis to the discussion of the collapse of a beam resting on a large number of equally spaced supports. Thus by the end of the eighteenth century, these ideas of collapse behaviour were available to the *École Polytechnique* and the *École des Ponts et Chaussées* – these schools were also fully conversant with Daniel Bernoulli's elastic equations of flexure, with Euler's buckling analysis and with the strength-of-materials problem of stresses due to bending. As has been noted, the linear elastic distribution of stress had become accepted as the 'correct' solution to the bending problem – safety was to be assured by ensuring that the greatest stress did not exceed some proportion of the limiting strength of the material.

Thus it was not the collapse state that was of interest – collapse was, in any case, hypothetical, since it would occur at some overload incorporating a factor of safety. The engineer requires his buildings to stand up, not to fall down, and so it was the working state of a structure that had to be examined. This, then, became the formal problem to be solved in the field of the theory of structures – the determination of the actual state of a hyperstatic structure under the action of specified working loads.

Navier 1826

As was noted in Chapter 4, Navier had attended both the *Polytechnique* and the *Ponts et Chaussées*, and so had full access to all the scientific work housed in those schools. He was appointed professor at the *Ponts et Chaussées* in 1830, but he had taught there for a number of years, and had already published his lecture notes as the *Leçons* of 1826. The books give an integrated account of the knowledge required by the civil and structural engineer – the second volume, for example, discusses problems in fluids and in the design of machine elements.

It is the first volume that deals with structural engineering problems; in particular, the fourth section exposes the 'Navier' philosophy of design that was to dominate the activities of structural engineers for over a century. Before this fourth section, Navier discusses the aspects of strength of materials that were described in Chapter 4 (it was this first section alone that was expanded at enormous length by Saint-Venant in 1864);

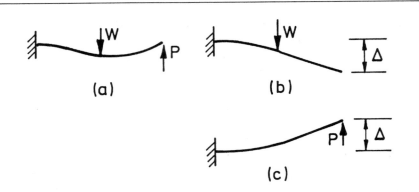

The cantilever beam in Fig. (a) is supported by a rigid prop at its right hand end, and is hyperstatic. If the prop were removed, as in Fig. (b), the beam would be statically determinate, and it would deflect at its tip under the load W by a certain amount, say Δ. The situation of Fig. (a) would be recovered if the value P of the propping force were such that, when acting alone on the freely cantilevered beam, Fig. (c), the tip deflexion had the same value Δ. Thus, in order to solve the deceptively simple problem of Fig. (a), an elastic flexural analysis must be made of the beam; the actual deflexions, greatly exaggerated in the figures, will be very small if the beam is to represent a realistic structural member.

the second section discusses problems of geotechnical engineering; and the third section is devoted to masonry arches.

The fourth section of Navier's first volume is concerned with the behaviour and design of timber structures. Beams in 1826 were, in general, made of wood, and Navier's first section, on the calculation of stresses, assumes that the internal structural forces in such beams are known. Navier's concern in the fourth section is with the way those internal forces may be calculated, and he takes as his example a modified Galileo's cantilever for

which the equations of statics no longer suffice. He imagines the cantilever
to be loaded not at its tip, but at some point within the length – the tip of
Galileo's cantilever will deflect slightly under the application of this load.
However, Navier's modified cantilever is prevented from deflecting at the
tip – there is a rigid prop at that point. If the value of the induced propping
force were known, then the internal forces throughout the beam could be
calculated – but how is the propping force to be found?

Navier saw clearly how a solution to this problem could be devised. The
statical equations of equilibrium must still hold, but no longer suffice by
themselves to determine the solution; the propping force is induced by the
tendency of the beam to bend under load; therefore some description of
the deflected shape of the beam must be introduced into the analysis. All
the scientific apparatus was available for this analysis – James Bernoulli
had stated that the elastic curvature of the beam was proportional to the
value of the bending moment at each point, and Daniel Bernoulli had shown
how to set up the differential equations of elastic bending to determine the
deflected shape of the beam. Thus Navier could write these equations, and
find the solution such that both deflexion and slope were zero at the root
of the cantilever, and the deflexion was also zero at the tip. From this
analysis emerged at once the value of the propping force at the tip, and so
the actual state of the beam had been found.

The sophistication of this analysis may be appreciated if the calculations
are dissected into their constituent parts. Three separate kinds of statement
must be made for the solution of a hyperstatic structure:

(a) Equations of equilibrium are written – internal and external forces
acting on the structure must obey the laws of statics (e.g. Newton's
law that action and reaction are equal and opposite). If these
equations suffice to determine the action of a structure, then the
structure is statically determinate; otherwise, two further statements must be made.

(b) Elastic equations are written expressing the internal actions of a
structure. Thus, in a beam, a bending moment induces a proportional curvature; in a straight member under tension, the extension
will be proportional to the value of the tensile force; and so on.
Thus an accurate description may be made of the slight deformations that will be experienced by a structure in response to its
internal forces.

(c) These small deformations must satisfy certain geometrical statements of 'compatibility' – the members of the structure must fit
together before and after deformation, and the deformations must
be such that they satisfy any external constraints on the structure.

Thus, for Navier's propped cantilever, (a) the equations of equilibrium
may be used to find the internal forces in the beam, but only in terms of an
unknown propping force at the tip; (b) the elastic equations may then be
used to find the deformed shape of the beam, still in terms of an unknown
propping force; and (c) the requirement that the beam must have zero
deflexion at both ends, and zero slope at the root, determines the correct
value of the propping force.

Trajan's bridge over the Danube, built by Appollodorus of Damascus,
about 100 AD (detail from Trajan's column). The bridge had 20
masonry piers, and the timber arch plus truss construction was in
timber with spans of over 30 m.

Navier repeats this formal plan of attack in his discussion of the trussed structure, composed of straight bars working (nominally) in pure tension or compression only. It had been known from ancient times that three bars, pinned together at their ends to form a triangle, would provide a stiff plane structure. Such triangulation can be extended by the addition of further members to create a large span bridge; the bas-reliefs on Trajan's column, 114 AD, show such construction in the timber bridge over the Danube. Much later, Palladio gave designs for different types of timber truss bridges for spans, typically, of about 30 m, and by the end of the eighteenth century such bridges were being constructed with spans of over 100 m. Cast iron and wrought iron began to be used for trusses – in the United States for railway bridges from about 1840 – but metal construction was not common when Navier wrote in 1826.

Navier exposes his analysis in terms of a simple two-dimensional wooden truss with one extra member – two bars connected to the ground would give a simply stiff framework, and one extra bar makes the truss hyperstatic. (Navier notes that his method will serve equally for the three-dimensional truss; three bars – the tripod – will be statically determinate, and the addition of an extra support again produces a hyperstatic structure.) Thus when the equations of statics are written, they do not suffice to determine the values of the forces in the bars, and the other two sets of equations must enter the analysis. The joints of the truss are imagined to be displaced by small amounts, unknown for the moment – the elongation of each of the bars can then be determined in terms of these displacements so that the bars still fit together; the joint displacements must be compatible with the bar elongations. Finally, the elastic equations relate the force in a bar to its extension, and enough equations can be written to solve the problem.

The formal procedure is completely straightforward, but, for two reasons, the actual calculations can be prohibitively heavy. First (for the truss problem, but in general for any but the most trivial structure), the analysis of the geometry of deformation is very complex. Second, although ways can be found to simplify this analysis, as noted below, the number of equations to be solved is very large. With the electronic computer not yet invented, much effort was spent in the nineteenth century on the derivation of techniques which would bring the labour within manageable compass.

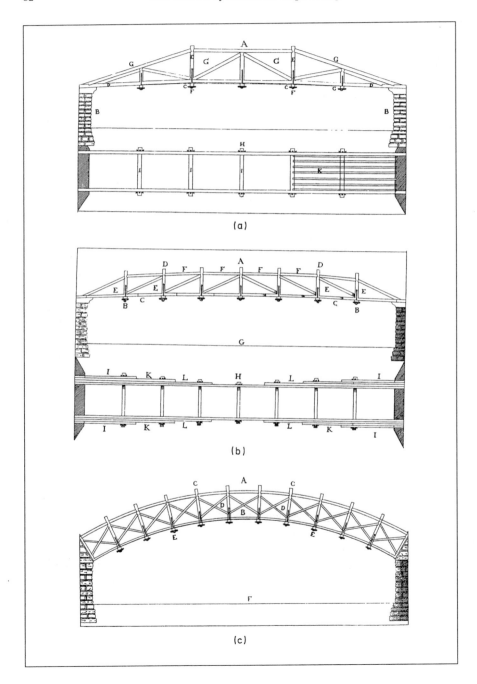

(a)

(b)

(c)

Palladio's *Four Books of Architecture* were published in Venice in 1570, and were at once enormously influential; Palladianism was embraced throughout Europe, and arrived in England through Inigo Jones (for example, the Banqueting House in Whitehall of 1620). The layout of the four books recalls Vitruvius – Book 1 deals with materials, techniques and the Orders; Book 2 with private houses (the "Villas"); and Book 4 with Roman temples. Included in Book 3 are designs for bridges.

Palladio was fully aware of the stiffness that can be achieved in timber structures by means of triangulation, and he illustrates the bridge at Cismon (no longer existing), Fig. (a), and gives also three other "inventions", of which two are shown in Figs (b) and (c). The construction of these bridges is extraordinarily ingenious, and is described carefully by Palladio, although he gives no details of the falsework or scaffolding that might have been needed – it must be assumed that the members are supported as required until the bridge is complete. In Fig. (a), the cross-members I, the bearers, are first placed – these will finally carry the deck K of the bridge. The long lower booms D, spanning from shore to shore, are next placed on the bearers I. The posts E are then erected on top of the booms, and an iron strap is bolted to the side of each of these; the strap passes through a hole in the bearers, and is secured on the bottom at F. The bridge is completed by adding the top boom GAG and the diagonal bracing, at which point any temporary supporting structure can be taken away. Construction in Figs (b) and (c) follows the same sequence.

Figure (a), at least, was based upon the construction of a real bridge, even if Palladio's other three designs are theoretical. There are two features, however, which make it seem unlikely that Palladio himself had been deeply engaged in bridge construction. The joints in the top boom GAG in Fig. (a) indicate, correctly, that the boom acts in compression. Correspondingly, the bottom boom at roadway level is in tension – the span is stated to be 100 ft, but there is no discussion of the very difficult problem of connecting wooden baulks to make an effective tension member of this length. Secondly, although the form of truss in Fig. (b) is highly efficient, the disposal of the material is precisely wrong – the lower boom should be strengthened towards the centre and not towards the ends.

Figure (c) gives a very effective arrangement of timber members to simulate, and in fact to act structurally as, a masonry arch.

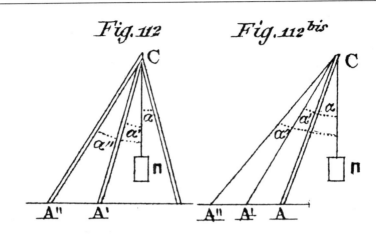

Navier's illustrations from his 1826 *Leçons* for the discussion of the hyperstatic truss. The members of the truss lie all in one plane, and are fixed to the ground at points A etc; they meet at the point C, where they are pinned together. In the left hand picture it is to be expected that all the bars will be compressed, while on the right two might act as ties. Navier does not discuss this – in either configuration of the truss only two equations of equilibrium are available, and the values of three bar forces have to be determined. Navier shows how the use of other structural equations lead to an elastic solution.

Navier does not discuss the fact that unless the three bars of the truss are made of exactly the right lengths, they will have to be strained when they are assembled together at the joint C. The ability to be in a state of stress in the absence of any external load is a characteristic of the hyperstatic structure.

Simplifying the working

The equations of statics for trusses are much easier to handle than the geometrical equations. James Clerk Maxwell, who worked in many fields and not only in the study of electromagnetic waves, made an extraordinary contribution to the truss problem in 1864. In fact, at the age of 19 (in 1850), Maxwell had written a massive paper 'on the equilibrium of elastic solids', and had covered a very wide range of problems, from the bending and torsion of beams to the centrifugal stresses in a rotating disk, and he continued to work throughout his life in the field of solid mechanics as well as those of heat, electricity and magnetism. In 1864, in a lucid but austere paper, without illustrations, he showed that a problem in the geometry of deformation could be replaced by a much simpler problem in the statics of forces.

From this work stems directly Maxwell's Reciprocal Theorem. This theorem seems to be most unlikely, but it unlocks one door to the analysis of elastic structures. In summary (and in a style reminiscent of Maxwell's own), if a known force is applied at point A of an elastic structure and, as a consequence, a resulting deflexion occurs at point B, then if the known force is applied at B, the *same* deflexion will occur at A.

A broader statement of the Reciprocal Theorem was formulated later, and independently, by Betti in 1872, and this led eventually, in the twentieth century, to an experimental technique for providing solutions to the elastic equations. As has been seen, small-scale model tests had been made as early as the seventeenth century, although structural engineers seemed to have preferred analysis to experiment. However, by the middle and late nineteenth century, important experimental results were obtained for problems where theory either did not exist or had been insufficiently developed – for example, for complex problems of buckling. All of these tests were *direct*; that is, a full scale replica, or a model, was made of the structural element concerned, and this was then tested to see how it behaved under load. Some models were very ingenious, using for example cardboard instead of iron in order to make easier observations – scaling laws and other aspects of the theory of models were well understood. No one concerned themselves with direct tests on structures – trusses and beams, for example – for which Navier's elastic theory had been fully worked out.

Indirect model tests, however, are designed to provide solutions to the mathematical equations, and do not attempt to reproduce the actual

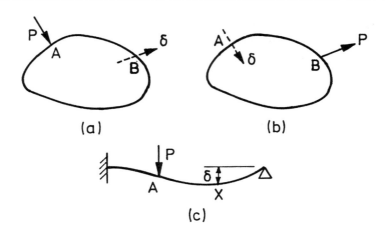

Maxwell's Reciprocal Theorem. Figure (a) represents an elastic structure, which might be a steel lattice truss, a reinforced-concrete beam or a masonry arch. A force of known value P acts in a known direction at point A, and as a consequence a deflexion δ occurs at point B in a certain direction. In Fig. (b) the force P is now applied at B; the same value of deflexion will be measured at A.

In Fig. (c) is shown a propped cantilever – the hyperstatic beam analysed by Navier. A load P on the beam will produce an elastic deflexion δ at some point X – the diagram shows the complete deflected form of the beam. The Reciprocal Theorem states that if the load P is applied at X, the same deflexion δ will be measured at A. The diagram can therefore be interpreted as an *elastic influence line*; it is a graph of the *deflexion at a fixed point A* as the load travels from one end to the other of the beam.

behaviour of the real structure. Instead, they make use of the Maxwell/ Betti Reciprocal Theorems. A proper model is made which reproduces from section to section the elastic properties of the real structure. Apart from this requirement, any material can be used, and Beggs, in 1927, proposed celluloid – stiff cardboard can be used to give less accurate but still acceptable results. Instead of loading the model, it is *deformed* in a specified way, and consequent displacements of any required sections are recorded. These displacements are transformed by the reciprocal theorem into forces, and simple experiments can determine the internal forces in a structure which result from the application of a specified loading system. In this way, all the heavy mathematics is avoided – the structure itself, or rather the model of the structure, carries out the necessary analysis.

Other ways were devised in the second half of the nineteenth century to reduce the heavy labour involved in the solution of large numbers of equations. Some of these ways were purely mathematical – numerical techniques were evolved in which approximate solutions could be obtained at a fraction of the labour cost of an exact analysis, and, moreover, the approximations could be made as accurately as required. Similarly (and in a way to be followed in the twentieth century by the construction of indirect models), the mathematics could be represented on the drawing board, and the science of "graphic statics" was developed in Switzerland and Germany, and held sway for the best part of a century as part of the tool kit of the structural engineer. Once again, Maxwell contributed to the mathematics which lay behind the construction of these diagrams, but it fell to the engineers to apply the work meaningfully to the solution of practical problems.

Maxwell was also aware of the implications of stored energy in elastic structures, although he did not pursue these ideas. It was noted that the Bernoullis, and Euler, had used the notion of strain energy in bending to solve the problem of the shape of the elastica – the calculus of variations determined the form of the bent strip such that the energy stored was a minimum. There had been some further English work on this matter in the mid-nineteenth century, but Castigliano, in 1879, gave a full formulation of the use of energy principles to find solutions for elastic structures. Castigliano developed the theorems which bear his name with reference to trussed structures, but he himself extended the work to beams and to masonry arches.

Very simply, if a structure is hyperstatic, and therefore has many ways – in fact, an infinite number of ways – of carrying specified applied loads, then the actual elastic state of the structure will be such that the strain energy stored is a minimum. It is, of course, on the basis of this theorem of Castigliano, easy to imagine that the structure is in some sense alive, and that it chooses a state in which it is, on average, and to speak loosely, least stressed. The theorem was in fact Castigliano's second; the first, from which the second derives, connects loads and displacements of an elastic structure by examination of the stored strain energy.

Use of energy in this way leads to precisely those equations that may be found directly from Navier's approach – once again, however, the analyst does not have to study the complex geometry of deformation in order to derive the equations. One difficult part of the analysis is avoided, but the number of equations to be solved still remains very large. Thus, by say 1880, efficient methods were available to formulate the equations governing the elastic behaviour of structures and, as has been seen, attention then moved to the question of how to solve those equations, whether by approximate numerical methods, by graphical representation or by indirect tests on models. Theory was not complete – for example, difficult problems of elastic buckling still awaited solution – but theoretical work in the first half of the twentieth century was devoted mainly to refining and simplifying the approximate methods.

These developments were finally terminated in the 1950s, with the invention of the electronic computer. The mathematics of elastic structural analysis had already been formulated in compact terms – the nineteenth-century equations could be written, for example, in terms of matrices. Now these vast arrays of numerical information could be handled by the computer, and the nineteenth-century equations could be solved exactly, or at least with as close an approximation as engineers wished. The engineer had merely to describe the structure to the computer – its overall geometry and points of attachment to foundations, the sizes of the members, the properties of the material, and so on – and to specify the loads to be carried; the computer would then print out values of the internal structural forces, and the corresponding stresses, to confirm the safety of the design. If the structure proved, in this analysis, to be unsatisfactory in any way, then the computer program could itself make changes to the design – by altering member sizes, for example – until the design criteria had been met.

No thought was given to the fact that, although the equations could at last be solved, the solutions might not in fact represent adequately, or indeed in any meaningful way, the real behaviour of a structure.

Chapter 7

Plastic Theory

For the development of a scientific theory, experiments are always useful in those areas where theory has not yet reached; they will indicate, perhaps, unforeseen behaviour which can then be incorporated in relevant analysis. Thus the design of connexions for steelwork or the disposition of reinforcement for concrete are difficult problems – amenable to analysis, but illuminated greatly by the results of practical tests. Indeed, progress in design in these areas would not really have been possible without the guidance of experiments.

There was no such compulsion to make experiments for mainstream structural analysis. Navier's formulation, exploited and perfected by nineteenth century engineers, appeared to be logically and self-evidently correct. Experiments to confirm the theory would be otiose – if there were disagreement between test and theory, this would indicate merely that the test had been badly made. Thus Mariotte's tests on fixed-ended beams in the seventeenth century – which led to the genuine and important observation that the breaking strength of a fixed-ended beam was twice that of the corresponding beam when simply supported – were not followed up during the next 200 years. Experiments were indeed made in this period on the strength-of-materials problem – the breaking strength of Galileo's cantilever beam – but there are only isolated instances of the application of these results to hyperstatic structures. In any case, after 1826 the breaking strength of a structure was not the objective of analysis; Navier had stated clearly that the maximum stress in a structure was not to exceed a certain fraction of the limiting stress of the material, and structural analysis was concerned with the identification of the most critical section of the structure and the calculation of elastic stress at that section.

For example, simple 'Navier' theory concludes that the bending moments (and hence the stresses) at the ends of a fixed-ended beam under uniform loading have twice the value of the bending moment at the centre of the beam. Thus the ends of the beam are critical for design. It was in 1914 that Kazinczy, in Hungary, tested steel beams whose ends were embedded in substantial abutments. The experiments were not made to prove or disprove the theory, but were directed to a practical end – Kazinczy wished to know whether the embedment could be taken as complete and, if not, what degree of fixity might be assumed.

As the load was steadily increased on a test beam, it was seen that, as predicted, yield occurred first at the ends. However, the beam could go on to carry further load, and it was not until a substantially greater weight had been added that deflexions became very large. When unloaded, each beam was found to have permanent kinking deformation, at the two ends and at the centre. Kazinczy called these 'kinks' hinges, and he stated that a fixed-ended beam could not collapse – that is, undergo largely unrestrained deflexions – until three hinges had formed. The two end hinges had merely turned a fixed-ended beam into one whose ends were still supported, and the beam was still viable as a structure; the third central hinge was necessary if collapse was to occur. Moreover, observed Kazinczy, the degree of end clamping to ensure the attainment of this collapse state is irrelevant, provided the embedment is strong enough to allow the hinges to develop. Thus a crucial, if surprising, answer was given to the question for which the experiments had been devised.

As a result of these experimental observations, Kazinczy deduced at once that the strength of a 'fixed-ended' beam was always twice that of the corresponding simply-supported beam. Moreover – and this turns out to be one of the most significant observations – the ends of the beam are not 'fixed' in the Navier sense; that is, the slopes of the ends of the beam are not exactly zero. Rather, this precise geometrical restriction is replaced by the loose statement that the end restraints must be sufficiently strong for the full strength of the beam to be developed.

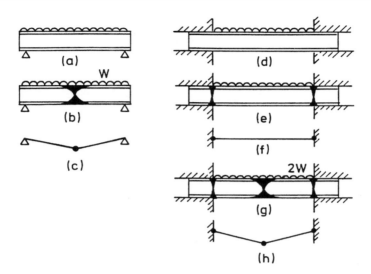

In Fig. (a), a steel I-beam is simply supported at either end, and carries a load uniformly distributed along its length. As the intensity of this load is slowly increased, the material near the centre of the beam yields, and passes out of the elastic into the plastic range, Fig. (b). When the central section is fully plastic, a *plastic hinge* has formed, and the *mechanism*, shown schematically in Fig. (c), will permit unrestrained deflexion at the collapse load W.

Figure (d) shows the same I-section with its ends now embedded in strong abutments. As the load is slowly increased, plastic hinges form at the ends, as in Fig. (e); however, Fig. (f) does not represent a collapse mechanism. The load must be increased to the value of $2W$, shown in Fig. (g), for the mechanism of Fig. (h) to develop.

The collapse strength of a fixed-ended beam is always twice that of the corresponding beam with simple supports, whatever the specified loading.

The 1936 Berlin Congress

The study of structures which, when loaded, pass out of the elastic into the 'plastic' range, was interrupted by World War I, but was resumed in the 1920s, mainly in central Europe – in Germany, Poland and Austria, and also in Switzerland and France. Structural engineers had established an international forum, the International Association for Bridge and Structural Engineering; the first Congress in Paris in 1932 was succeeded by a second in Berlin in 1936, at which eight papers on plasticity theory formed one section of the proceedings. These papers were concerned with the collapse state of structures, that is, with their ultimate strength; just as Galileo had been concerned with the breaking strength of a statically determinate cantilever, so now the topic was the strength of hyperstatic structures. In general, the structures under discussion in Berlin in 1936 were examined in a 'historical' way – that is, the 'Navier' elastic solution was first obtained, and this solution was modified to allow for plastic behaviour as the loading was increased towards collapse. It is, of course, not surprising that engineers in the 1930s should start from a conventional elastic approach in attempts to describe the actual behaviour of structures. Kazinczy himself attended the Congress 20 years after his crucial experiments, and he advocated such an approach.

The results of one of the important experimental papers, by Maier-Leibnitz, were interpreted by the author in just this way. Maier-Leibnitz tackled experimentally the simplest hyperstatic problem, the beam on three supports – Galileo's marble column with a central prop, or Navier's own analysis of the continuous beam. In one series of three tests, Maier-Leibnitz first set the three supports at the same level before loading the beam to collapse; in the second test, he lowered the central support slightly before loading; and in the third test he raised the central support. The 'Navier' solutions for each of these three configurations differ widely from each other – indeed, in the third test Maier-Leibnitz placed the central support at such a height that yield was just occurring there, so that the 'Navier' philosophy precluded the application of any load to the beam. In fact, Maier-Leibnitz found that the collapse loads for each of the three beams had the same value.

He showed easily how this apparently anomalous result came about. As the loading was increased on the beam, and the material passed out of the elastic into the plastic range, so the original elastic distribution of stress was modified until, at the collapse state, those plastic hinges necessary for collapse were formed (just as Kazinczy had found 20 years earlier that plastic hinges at the ends and centre of a fixed-ended beam were necessary for collapse). The final strength of a ductile structure was dictated by its plastic behaviour, and was not affected by accidental or deliberate distortion, or by any other imperfection of construction.

In a theoretical paper of great importance, F. Bleich pursued this question of imperfections, such as settlement of foundations, forcing together of members because of slightly incorrect manufacture, and so on. He notes that such imperfections can leave a hyperstatic structure in a state of stress, even without the application of external loading. Each member has only a limited capacity, however, before yielding occurs; if it starts from an initially stressed state, it will yield earlier or later than otherwise, but will finally reach the same limiting state – initial elastic stresses are 'wiped out' as the material becomes plastic. The three plastic hinges in Kazinczy's (or Maier-Leibnitz's) beams, with known values of full plastic moment, are the sole determinants of the final overall strength. (Bleich notes that differential temperatures between various parts of a structure can lead to states of self-stress similar to those induced by physical imperfections, and once again these cannot influence the final strength of a structure.)

This discussion of self-stress in fact comes after Bleich has made a number of statements crucial for plastic methods of structural design. First, he abandons the idea of a factor of safety on the calculated value of elastic stress. Instead of determining the most critical section of a structure in its working state, and basing the whole design on the stress at that critical section, Bleich examines the hypothetical collapse of the structure under increased external loading. He introduces the idea of a *load factor*, defined as the ratio of the collapse load to the specified design working load. That is, the working loads on a structure (self-weight, floor loads, wind pressure and so on) are imagined all to be increased in proportion until sufficient plasticity has developed for large deflexions to occur – such as the large deflexions which developed at collapse of Kazinczy's beams. In fact, Bleich made an alternative formulation of the load factor. The loads are not imagined to be increased; instead, the design is carried out at the specified working values of the loads, but with a material whose yield stress is *reduced* by the value of the load factor. Thus for a load factor of 2, say, the structure is *designed* to collapse under the given loads, but is actually *built* twice as strong as this calculated design.

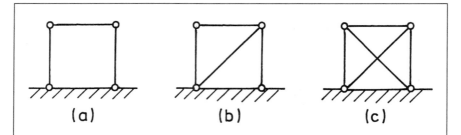

The figures show plane, two-dimensional trusses. Figure (a) does not represent a structure – the three bars, pinned freely together at their ends and to the foundation, form a mechanism incapable of carrying any load. The addition of an extra bar, Fig. (b), produces a stiff tri-angulated structure. If any one of the members is manufactured with a slightly incorrect length, the structure may still be freely assembled without strain – the geometry will then be inexact, but the eye will not notice this. By contrast, the truss of Fig. (c) is hyperstatic; if any one of the bars has the wrong length, then the truss, when assembled, will be in a state in which all the bars are strained. This ability to experience a state of self-stress in the absence of external loading is characteristic of hyperstatic structures.

Second, Bleich defines the characteristics of the material needed for a plastic theory of design – if the method is to be valid, then an elastic stress-strain relationship must be followed by indefinitely large deformations at the plastic limit. 'Large' in this context is a relative rather than an absolute term. It was noted that a bar of mild steel 1 metre long would reach its yield stress at an extension of about 1mm; fracture of the bar may not occur until the extension is about 250 mm or more. Adequate ductility will be present in practice if the extension at fracture is an order of magnitude less – say 25 mm. The material must, in fact, be ductile, in the way that constructional mild steel, or wrought iron, is ductile, and in the way that reinforced concrete can suffer relatively large deformations without excessive distress. Brittle substances, such as cast iron or glass, on the other hand, are poor structural materials; if it were possible to assemble glass members into a building frame without breakage, the structure would certainly shatter catastrophically on first loading.

Finally Bleich cuts loose completely from conventional elastic calculations. He shows that it is not necessary to compute the elastic behaviour of a structure under a set of specified loads, and then to modify this solution as the loads are (hypothetically) increased to the collapse state. The equilibrium equations by themselves admit of infinitely many solutions – Barlow's model voussoir arch is stable under widely varying positions of the thrust line. Of these many solutions, the Navier, elastic, solution also satisfies both the stress-strain relationship and the boundary conditions – it is the joint satisfaction of these three types of equation that makes the elastic solution so complex. However, Bleich showed that any one of the equilibrium solutions, easy to construct, could be used as a starting point for the plastic calculations; the collapse state could be considered directly, without examination of an imaginary increase of the working loads.

J.F. Baker

The 1936 Congress in Berlin was held against the background of the growing use of steel for industrial and large commercial and domestic buildings. Steel frames had been in existence since the start of the twentieth century; they were designed on the basis of elastic theory, embodied in codes of practice for the benefit of the hard-pressed engineer. There were many such codes in use throughout the world and, while these codes were agreed on basic principles, they differed on almost every point of detail. As a single example, the building regulations in the United Kingdom were in the hands of the local authorities, and different floor loads were prescribed in Edinburgh, Glasgow and Newcastle, while the London County Council had yet another range of values.

In 1929, the British steel industry set up the Steel Structures Research Committee to try to bring some order into practical steel design. The SSRC included in its membership eminent academics and officers from Government research stations, and leading representatives from the consulting and contracting professions. The Committee appointed J F Baker (later Sir John Baker, and then Lord Baker of Windrush) as its full-time technical officer; it was his task to assemble technical information, to write or commission papers developing theory, and to oversee the collection of experimental evidence. The findings of the Committee are collected as papers in three volumes published in 1931, 1934 and 1936.

These papers reveal the state of understanding, in the UK in 1936, of the structural design process, as indeed does Baker's own textbook (written with A J Sutton Pippard) published in that same year. Design and analysis are completely elastic processes, and the tools available include, for example, Maxwell's reciprocal theorem and the energy theorems of Castigliano. All of this is nineteenth-century analysis, although an exception lies, perhaps, in the treatment of beam-columns – that is, of members liable to buckle in an 'Euler' way under axial loading but subject also to bending. Theoretical advances are made in the solution of this problem, which is important for columns in steel-framed buildings.

However, the outstanding contribution of the Steel Structures Research Committee to the problem of structural design lay in experimental work. New steel buildings were being constructed in the 1930s, and the Committee arranged for tests to be made on several structures, among them a nine-storey hotel block, an office building and a block of residential apartments. For the first time, stresses in real structures were measured – the development of suitable strain gauges was an essential part of the work. The results were published in detail and were given extensive analysis, but they can be summarised easily – the real stresses measured in the buildings under test bore almost no relation to the stresses calculated by designers using the available elastic methods.

The SSRC was not slow to find the reason – small, and unpredictable, errors of manufacture and erection were enough to invalidate almost completely the elastic calculations, which are extraordinarily sensitive to small imperfections of geometry or lack of fit. The SSRC did its best to devise some rules for design – allowing, for example, for the flexibility of connexions between the steel members – but Baker knew, in 1936 when the Final Recommendations for Design were published, that these were deeply flawed. Moreover, the rules applied only to the simplest array of beams and columns; nothing like general recommendations had been given which could apply to more complex building structures. In a sense, Baker's 1936 text with Pippard is a final exposition of structural analysis as it was known at that time; it offered no possibility of advance or development, and it seems now not to be relevant to the purpose of structural design.

With no appreciation of the fact that a structure might contain imperfections of one sort or another – indeed, with no way of predicting what
those imperfections might be – the elastic designer was making the assumption that the structure was perfect; that is, that the supports were rigid,
the beams were fitted exactly to the columns, and so on. In the case of
Navier's propped cantilever, the elastic designer assumes that the clamp
at one end is perfectly rigid, and that the prop at the other end stays at
exactly the same level. Very small displacements at either end of the beam
will totally alter the elastic stress distribution.

The elastic calculations refer to the perfect structure, and not to any
real construction. However, although the measured response of a steel
frame is very far from that predicted, the elastic designer's assumptions
do seem reasonable; common sense would support the belief that a trivial
defect cannot really affect the strength of a structure. Common sense is in
this instance correct, and the paradox is resolved by concluding that the
calculation of elastic stresses is not relevant to the prediction of strength.
The strength of a real structure made of ductile material does not depend on
an elastic stress reaching some limit at one point in the structure; it is given
by the steady development of unacceptably large deformations. Navier's
methodology, formulated over a hundred years earlier and developed over
the century into a coherent body of scientific knowledge, was, it turns out,
not directed to the solution of the practical design problem.

The ultimate strength of structures was, of course, precisely the study
reported in the section on plasticity of the 1936 Berlin Congress of the
International Association for Bridge and Structural Engineering. As has
been noted, Maier-Leibnitz, and Bleich and others at the Congress, were
aware that small imperfections were of no consequence to ultimate
strength. Baker went to Germany in the wake of the Congress, and he met
Maier-Leibnitz, from whom he learned, for the first time, of plasticity, and
of the experimental result that collapse loads of continuous beams were
virtually unaffected by misalignment of supports, that is, by the type of
imperfection identified by the SSRC. Baker saw at once that the way forward in structural design lay in the exploitation of plastic ideas.

He set up an intensive investigation into the plastic behaviour of steel
structures, first in Bristol in 1936, where he had been apppointed professor,
and from 1943 in Cambridge. Within the astonishingly short period of
a decade the work had borne fruit – in 1948, the British Standard 449
(The Use of Structural Steel in Building) was altered by the insertion of a

single clause permitting plastic design. (BS 449 was essentially the elastic design code put forward by the Steel Structures Research Committee as an interim measure at the start of its work in the 1930s.)

Baker's approach was, above all, experimental. He repeated Maier-Leibnitz's tests on continuous beams, and made the first substantial series of tests on portal frames (and later on multi-storey structures). A spectacular application of the theory was to the design of the Morrison shelter, which was installed in over a million households in the UK in World War II. The intellectual basis of the design was both simple and elegant, and unattainable, indeed unthinkable, by the conventional elastic designer. The shelter, of the shape and size of a dining table, under which the family could sleep, was required to squash down plastically by no more than 12 inches if the house collapsed. The energy released in the collapse of a house can be estimated with some accuracy; the energy absorbed in the plastic deformation of steel can be calculated almost exactly. Equating the two gives the design of the shelter.

Despite this energy calculation, the plastic analysis of beams and frames was tackled by Baker as a problem in statics. For fairly simple structures, the collapse mechanism was known, or could be guessed accurately – Kazinczy's fixed-ended beams could collapse only by the formation of hinges at their ends and centres. The knowledge of the value of the moment of resistance at these hinges immediately unlocks the statics of the problem – a hyperstatic structure becomes statically determinate at collapse. At a supposedly clamped end, for example, the elastic analyst would assume zero slope and zero deflexion of the beam. These geometrical restraints are virtually impossible to satisfy in practice – they represent the 'imperfections' which make the elastic solution such a poor description of actual behaviour. At collapse, however, an unattainable condition of zero slope is replaced by a precise numerical condition of statics – the value of the plastic moment of resistance is entered into the equations.

This kind of statical approach to the plastic analysis of structures, without any real mathematical principles to underpin the theory, was adequate for the design of even quite complex steel buildings. A host of ancillary problems had to be tackled, and they were tackled successfully, at least from a practical point of view – the effect of shear force and axial load on the formation of plastic hinges, for example, or the behaviour of columns tending to buckle in the elastic/plastic range. It was an extraordinary achievement to have moved, within a little over 10 years, from an idea glimpsed by Baker in 1936 to an officially permitted design method. However, each structure, or structural type, was tackled *sui generis*; Baker, in 1948, did not yet know of the fundamental plastic theorems.

The 'safe' theorem

Attempts to establish basic principles are evident in the papers of the 1936 Berlin Congress. However, World War II was very close, and the extreme unrest of the times put a brake on further progress. Several papers at the Congress had referred to the work of W. Prager, although he himself had not made a contribution to the proceedings. Prager was one of those who left Germany shortly afterwards, and in 1941 he established a Division of Applied Mathematics at Brown University in the United States, which was to make outstanding theoretical advances in plasticity theory (and in other areas). In the same year of 1936, a Russian, A.A. Gvozdev, presented a paper to a conference, which was not in fact published in Moscow/Leningrad by the Academy until 1938; the paper was unknown outside Russia, and little noticed within that country. This work became known to Prager about 1948, together with other statements of the principles of plasticity theory, and he and his colleagues supplied proofs of the fundamental theorems in 1949, with full acknowledgement to the original Russian scientists.

The mathematical theorems stated and proved by Gvozdev provide the rigorous backing for Baker's advances in engineering plasticity. Gvozdev is clear, first of all, on the assumptions on which the theory rests. Paramount among these is the requirement that the material should be ductile – a requirement being enunciated simultaneously but independently in Berlin by Bleich. Increasing deformation of a structure might be accompanied initially by a proportional increase in stress, but a limit is reached at which indefinite increase in strain can occur without any fall off in the corresponding resistance of the material. Most importantly, ductility must be evidenced not only locally, at individual portions of the structures, but also overall – and this implies that the basic plastic theorems to be proved by Gvozdev must be applied with caution if there is any danger of instability. For example, a buckling column cannot sustain the external loads causing the instability; as soon as the column displaces, the loads fall dramatically, and this structural element, even if made of mild steel, is effectively 'brittle' rather than 'ductile'.

If, however, the overall structure and its elements satisfy the prime requirement of ductility, then Gvozdev notes that only three types of equation may be written. (These echo closely the statements by Navier in the context of the elastic theory of structures.)

First, the equations of equilibrium must be satisfied; internal structural forces must balance the externally applied loads. There are in general an infinite number of solutions of these equations for the hyperstatic structure; only if the structure is statically determinate is the solution unique.

Second, the yield condition must be satisfied; none of the internal stresses must exceed the known yield limit of the material.

Third, there must exist some mechanism of deformation at collapse of the structure; for Kazinczy's fixed-ended beams, plastic hinges at the ends and centre will permit the development of large displacements, with correspondingly more complex mechanisms for more complex structures.

Gvozdev proved three theorems based upon the use of these three different conditions. The fundamental theorem is that of *uniqueness*; if all the conditions are satisfied simultaneously, then the collapse load corresponding to the solution of the equations has a definite and calculable value. Baker's engineering theory was based on solving the three types of equation simultaneously; Gvozdev's uniqueness theorem confirmed that the collapse loads calculated by Baker were correct, and that there was no possibility of alternative solutions leading to different values of those collapse loads.

It will have been noted that the three conditions laid down by Gvozdev make no reference to any initial state of the structure. The structure before loading could be in a state of self-stress due to those imperfections which have been mentioned – the forcing together of members during manufacture to rectify small dimensional errors; the settlement of supports or slight 'give' in abutments (as at the ends of supposedly fixed-ended beams); or, in a slightly different class but with the same effect, the firm bolting together of members at a connexion which was regarded by the designer as a single 'pin'. However they arise, such states of self stress cannot affect the unique and definite value of the collapse load.

A second theorem stated by Gvozdev has proved of great value in devising techniques of analysis of structures in order to determine their collapse loads – it is the *unsafe theorem*. If attention is concentrated on possible collapse mechanisms, and there is no requirement that equilibrium should be satisfied, and moreover the yield condition is not necessarily satisfied everywhere in the structure, then it is still possible to calculate a value of the collapse load. That value, however, is unsafe; the designer will believe the structure to be stronger than it is in reality.

The rock on which the whole theory of structural design is now seen to be based is provided by a third theorem of Gvozdev, the *safe theorem*. If the engineer can find a set of forces within the structure which equilibrate the external loads, and for which the yield condition is not violated (that is, Gvozdev's first two conditions are satisfied, but no attention is paid to the requirement that there be a mechanism of collapse), then the corresponding value of the load on the structure is a safe estimate of the collapse load. The structure can in reality carry more load than the value given by this kind of calculation.

The theorem can be put very simply. If the designer can find a way in which a structure is comfortable under the application of specified loads, then the structure is safe. The power of this statement lies in the fact that the designer need find only *one* such way; this may not be the way the structure actually chooses to be comfortable, but if the designer can find a way, then so can the structure. The designer has no obligation to try and find the actual state.

The theorem explains why Navier's methodology gives safe, but in general, uneconomical designs. An elastic solution may describe a state which is not observable in practice, as was found by the Steel Structures Research Committee in the 1930s, but the elastic solution is one, out of an infinite number of equilibrium solutions, with which the structure is comfortable.

Most importantly, the safe theorem, and ideas of plasticity theory generally, direct the engineer's attention to the questions about structural behaviour that can be answered. *Prima facie*, it seems reasonable to try to establish a theory that will determine how it is that a structure carries given design loads – how will an industrial building respond to wind forces, weights of equipment, crane loads and so on? Once the actual state of the structure has been determined, that is, that part of analysis has been completed which falls within the *theory of structures*, then attention can be given to local conditions; the *strength of materials* will establish the size of members at every section of the structure. However, all of the experimental work of this century has exposed the paradoxical fact that there is no actual state of a structure in response to given loading – or rather, there is, of course, a state here and now in which the structure finds itself, but that state is ephemeral. A slight lurch in a gale, a small settlement of a foundation, a slip in a connexion, a differential rise in temperature – all of these are almost trivial external influences, but they will produce enormous changes in the actual state of the structure.

Moreover, this kind of 'imperfection' is unknown and unpredictable. It has been the triumph of plastic ideas, observed by Kazinczy in 1914, supported by observations in Berlin in 1936 and justified mathematically by Gvozdev in that same year, and developed energetically by Baker over the next decade, that trivial imperfections can in fact have no influence on the strength of practical ductile building structures. The theory was developed largely with reference to steel construction; it is in fact a universal theory that can be applied to a structure made of any material that an engineer would consider suitable for use in building. It explains, for example, how it is that a Gothic cathedral (and a Greek temple or Roman aqueduct) continues to stand – stresses are low, and the plastic theorems translate simply into the requirement that forces should be contained within the masonry. This is a geometrical restriction – the shape of the structure must be correct. Once the shape was established and recorded, medieval rules of proportion ensured that the building would survive assault from a hostile environment.

A milkmaid weighing 600 N (about 61 kg, 135 lb, $9\frac{1}{2}$ stones) sits on a three-legged stool. For what force should each leg of the stool be designed? The stool is supposed to be symmetrical, the milkmaid sits in the centre of the seat, and so on. The answer is, of course, 200 N.

The same milkmaid now sits on a square stool with four legs, and again the stool and loading are symmetrical. For what force should each leg of the stool be designed? The answer of 150 N is not necessarily correct. A robust, nearly-rigid milking stool, standing on a firm shed floor, will rock; three of the legs will appear to make contact, supporting the weight of the milkmaid, but the fourth will be clear of the floor. If the clearance is only a fraction of a millimetre, then it is certain that the force in that leg is zero. Simple statics shows that the force in the leg diagonally opposite will also be zero, even if it appears to be touching the floor. The milkmaid is supported by the two other legs, each of which carries 300 N.

The stool may be imagined to be placed randomly, and there is no way of deciding *a priori* which legs will be in contact – *all* legs must therefore be designed to carry 300 N.

The four-legged stool is hyperstatic; only three equations of equilibrium may be written, and they will only determine the forces in a statically determinate three-legged stool. Navier showed a way to solve the four-legged problem. Elastic information must be introduced; the flexural properties of the seat of the stool must be specified, as must the axial compressibility of the legs – the problem becomes very complex. However, computers may be used to solve the difficult equations, and the forces in the four legs are found all to be 150 N. The computer has, of course, assumed (as has the engineer using the elastic-analysis program) that the floor is level and that all the legs are of the same length – indeed, no other assumption is possible for the elastic solution, since the way the stool is placed on a rough floor is unknowable. The Navier methodology does not give the "actual" forces in the legs.

If, however, the elastic design is accepted, then the stool will be manufactured with each leg designed to fail, in a ductile way, at a load of 150 N. A suitable load factor must be used, say a value of 3, and the legs of the stool will be manufactured with a strength three times that calculated. A loading history may be recorded for this stool having legs of strength 450 N as a laboratory test is performed in which a central point load is slowly increased from zero. Initially, in general, only 2 legs will carry the load; when the load has reached the value of 900 N, these two legs will begin to 'squash', allowing the other two legs to make contact. The load may then be increased further to 1800 N, when all four legs are carrying their 'squash loads' of 450 N each.

Thus, although the elastic analysis predicts leg loads all of value 150 N, an experiment under a total load of 600 N will record two legs with zero load and two legs carrying 300 N. Nevertheless, by the time the load has reached its full factored value of 3 × 600 N the leg loads have equalized – actual ductile, plastic, behaviour has come to the rescue of elastic analysis.

The rescue can take place only if the structure *is* ductile. An unstable element in a structure is 'brittle' – if one of the legs of the stool should buckle sideways instead of gently squashing, then the whole stool will collapse. Thus if the elastic analysis is used as the basis of design, each leg will be designed to just remain stable at a load of $3 \times 150 = 450$ N – the designer believes he is designing to a load factor of 3, so that the stool would take the weight of a man of 1800 N (184 kg, 405 lb, 29 stones). In practice, however, a load of 900 N will cause two opposite legs of a randomly-placed stool to be loaded to 450 N – these legs will buckle, and the actual load factor is only 900/600 or $1\frac{1}{2}$, which may well be dangerously low, since the stool would collapse under the weight of a 92 kg man.

The most critical problem in the whole of structural design is to determine the worst possible loading on any particular member (the flange of a steel girder, a column in a skyscraper, a stringer in the wing of an aircraft), and to proportion that member so that it is strong enough, and so that there is no possibility of buckling.

Index